SpringerBriefs in Physics

SpringerBriefs in Physics are a series of slim high-quality publications encompassing the entire spectrum of physics. Manuscripts for SpringerBriefs in Physics will be evaluated by Springer and by members of the Editorial Board. Proposals and other communication should be sent to your Publishing Editors at Springer.

Featuring compact volumes of 50 to 125 pages (approximately 20,000–45,000 words), Briefs are shorter than a conventional book but longer than a journal article. Thus, Briefs serve as timely, concise tools for students, researchers, and professionals.

Typical texts for publication might include:

- A snapshot review of the current state of a hot or emerging field
- A concise introduction to core concepts that students must understand in order to make independent contributions
- An extended research report giving more details and discussion than is possible in a conventional journal article
- A manual describing underlying principles and best practices for an experimental technique
- An essay exploring new ideas within physics, related philosophical issues, or broader topics such as science and society

Briefs allow authors to present their ideas and readers to absorb them with minimal time investment.

Briefs will be published as part of Springer's eBook collection, with millions of users worldwide. In addition, they will be available, just like other books, for individual print and electronic purchase.

Briefs are characterized by fast, global electronic dissemination, straightforward publishing agreements, easy-to-use manuscript preparation and formatting guidelines, and expedited production schedules. We aim for publication 8–12 weeks after acceptance.

More information about this series at http://www.springer.com/series/8902

V. T. Bachinskyi · O. Ya. Wanchulyak ·
A. G. Ushenko · Yu. A. Ushenko ·
A. V. Dubolazov · Igor Meglinski

Polarization Correlometry of Scattering Biological Tissues and Fluids

 Springer

V. T. Bachinskyi
Bukovinian State Medical University
Chernivtsi, Ukraine

A. G. Ushenko
Optics and Publishing Department
Chernivtsi National University
Chernivtsi, Ukraine

A. V. Dubolazov
Chernivtsi National University
Chernivtsi, Ukraine

O. Ya. Wanchulyak
Bukovinian State Medical University
Chernivtsi, Ukraine

Yu. A. Ushenko
Correlation Optics Department
Chernivtsi National University
Chernivtsi, Ukraine

Igor Meglinski
Opto-Electronics and Measurement
Techniques Laboratory
University of Oulu
Oulu, Finland

School of Engineering and Applied Science,
School of Life and Health Sciences
Aston University
Birmingham, UK

ISSN 2191-5423   ISSN 2191-5431   (electronic)
SpringerBriefs in Physics
ISBN 978-981-15-2627-5   ISBN 978-981-15-2628-2   (eBook)
https://doi.org/10.1007/978-981-15-2628-2

This Springer imprint is published by the registered company Springer Nature Singapore Pte Ltd.
The registered company address is: 152 Beach Road, #21-01/04 Gateway East, Singapore 189721, Singapore

Contents

Chapter 1
Introduction

1.1 Vector-Parametric Reconstruction of the Distributions of Birefringence and Dichroism of Networks of Biological Crystals

Among the extensively used traditional methods for studying the turbid tissue-like phase-inhomogeneous media, an important place belongs to matrix polarimetry—based diagnosis of optically anisotropic biological samples [1–6]. Recently sufficient progress is achieved in theoretical description and modeling of optical radiation propagation in scattering tissue-like phase-inhomogeneous medium and in polarimetric microscopic study of images of collagen fibrillar networks of once scattering histological sections of tissues of human organs [7–13]. The obtained information extends the functionality of traditional histological studies of the optically anisotropic component of the morphological structure of biological tissues. An important step in objectifying such studies is the use of quantitative estimates—the determination of the size and number of fibrils, the reconstruction of their angular distributions [14–16]. A generalization of such methods was the methods of statistical analysis of coordinate distributions of the magnitude of the matrix elements of weakly scattering histological sections of polycrystalline layers (samples) of human organs [17]. The correlation between the sets of statistical parameters that describe the maps of optical anisotropy of the polycrystalline component of biological layers are determined [18]. As a result, the actual parameters for oncological screening and characterization different anomaly of biological tissues were suggested. In fact, the most of real biological media are depolarizing incident laser light [19]. Therefore, the reliable correlation between the parameters of the optical anisotropy of the morphological structure of the biological tissues and the topological maps (distribution) of elements of the polarization matrix becomes invalid [20–22]. Thus, the search for new approaches to the development of methods and systems of polarimetry [23–26] of depolarizing layers of biological tissues is theoretically relevant and practically important. For such

© The Author(s), under exclusive licence to Springer Nature Singapore Pte Ltd. 2020
V. T. Bachinskyi et al., *Polarization Correlometry of Scattering Biological Tissues and Fluids*, SpringerBriefs in Physics, https://doi.org/10.1007/978-981-15-2628-2_1

objects, the inverse problem of polarization "introscopy" is not solved, namely reconstruction and quantitative evaluation of the polycrystalline structure of fibrillar networks. Current study is devoted to the analysis of the diagnostic effectiveness of differential polarization mapping techniques [27–34]. This direction of polarimetry is based on the representation of the matrix operator in the form of a superposition of differential matrices of the 1st (polarized component) and 2nd (depolarized component) orders. This approach involves the synthesis of polarization mapping and reconstruction of the coordinate distributions of birefringence and dichroism of histological sections of fibrillar biological tissues (normal and pathological chan-ged) of various human organs (myocardium, tissues of reproductive sphere of a woman).

Modern microscopic diagnosis of such changes does not provide a sufficient level of accuracy. Presently, in the tasks of forensic medicine, one of the most difficult is the identification of acute coronary syndrome. At the moment, the expressivity and accuracy (65–70%) of the detection of such a pathology are unsatisfactory [35–37]. The accuracy of the analogous diagnosis of the first stage of endometrial cancer is 75%, the second stage is 66%, and only in the third stage it is increased to 88% [38, 39]. At the preclinical stage of no less severe pathology of prolapse of the genitals, quantitative estimates of changes in the morphological structure of the connective tissue component of the uterine ligament are absent. Therefore, the relevance of creating a set of new polarimetric techniques and sys-tems for detecting and monitoring changes in the parameters of optical anisotropy, which are caused by early pathological and necrotic conditions, is not in doubt.

1.2 Polarization-Correlation Analysis of Images of Optically Anisotropic Biological Layers

Optical methods using polarized radiation have found various applications in the remote and non-destructive diagnostics of optically anisotropic manifestations of the morphological structure of light-scattering layers of biological tissues and liq-uids [1–13]. The use of laser coherent probes ensured high accuracy and sensitivity of laser polarimetry techniques in detecting changes in the polycrystalline structure of biological preparations [17]. On this basis, new possibilities were identified in the early and quantitative diagnosis of cancer of biological tissues [18, 40, 41]. An important place among them is occupied by a statistical analysis of the distributions of the azimuth and ellipticity of polarization-inhomogeneous images of a wide range of biological tissues [18, 19]. This quantitative approach made it possible to differential diagnosis of not only cancer, but also pre-cancerous changes in the polarization manifestations of optical anisotropy of fibrillar networks of structured biological tissues. As a result, diagnostically important relationships were found between changes in the statistical structure (magnitudes and ranges of changes in average, dispersion, asymmetry and excess) of polarization maps of the azimuth and ellipticity of microscopic images and distributions of phase anisotropy parameters (optical axis directions and phase shifts of fibrillar crystallites) of biological preparations [42, 43]. The analytical basis of quantitative laser polarimetry methods has become the "single-point" method—determining the average and standard deviation of the set of statistical moments of the 1st–4th orders that characterize the distribution of the azimuth and polarization ellipticity at the points of digital images of representative samples of the studied biological samples [17–19]. As a result, for optically thin layers of biological tissues (skin, uterine wall, prostate, rectum, etc.), quantitative criteria for the effective differentiation of "norm-pathology" were determined. At the same time, this approach, based on the technique of "single point" mapping, does not allow obtaining information about weak phase changes in fibrillar networks [14–16]. The source of such information is the topological structure ("inaccessible" for statistical analysis) of polarization-inhomogeneous object fields of biological tissues. From a diagnostic point of view, information of this kind is very important in detecting intermediate stages of pathological conditions for which variations in the parameters of structural (birefringence) anisotropy are insignificant. To obtain such topological information allowed a new, polarization-correlation approach [41, 44, 45]. Analytically, it is based on determining the degree of correlation matching between types and forms of polarization of various points of the object field of the studied sample of biological tissue. A quantitative measure of this correlation is the module of the complex degree of mutual polarization (CDMP) of the microscopic image. The magnitude of the CDMP is determined by the degree of correlation matching between the azimuth and ellipticity of polarization of various points of the object field of a phase-inhomogeneous object. Within the framework of laser polarimetry, this theoretical correlation approach was expanded to analyze the topological structure of polarization-inhomogeneous microscopic images of optically anisotropic layers of biological tissues. As a result, new, additional criteria for differential diagnosis of weak phase changes were obtained, which are due to the varying severity of oncological changes in the tissues of the female reproductive sphere [23–26, 35]. At the same time, in the field of development and creation of new techniques and systems for polarization correlometry, a number of unsolved theoretical, applied, and diagnostic problems remain:

i. *Theoretical*—the construction of a universal theory for describing the processes of formation of the polarization-correlation structure of object fields of optically anisotropic biological layers using the generalized formalism of the "two-point" Stokes vector [46, 47];

ii. *Applied*—the absence of a methods and algorithms for experimental measurement of coordinate distributions of parameters (module and phase) of the "two-point" Stokes vector;

iii. *Diagnostic*—determination of the set of relationships between the statistical parameters of the maps of the module and the phase of the "two-point" Stokes vector and maps of optical anisotropy of not only fibrillar, but also parenchymal biological tissues.

First of all, the relevance of solving these problems is due to the need to increase the sensitivity of laser polarimetry techniques to weak phase fluctuations of the polycrystalline structure of biological tissues in the early stages of the onset of pathology. On the other hand, most tissues of human organs are parenchymal (liver, spleen, pancreas, etc.) and have a weak level of birefringence and dichroism. In this regard, our monograph is devoted to the analysis of the possibilities of developing and increasing the diagnostic efficiency of polarization correlation methods by using the "two-point" formalism of the Stokes vector parameters in the theoretical description and experimental mapping of polarization-inhomogeneous object fields of preparations of such biological tissues.

1.3 Fundamentals of Stokes Polarimetry of the Optical Anisotropy

A well-known fact is the presence of structural anisotropy (linear birefringence) of fibrillar networks, which are formed by optically active protein complexes with circular birefringence [2, 7, 9]. The most common case for such optically anisotropic objects is the presence of multiple acts of volume scattering, which partially depolarize the laser radiation. To describe the processes of formation of a polarization-inhomogeneous field of objects of this type, a layered matrix model is used [22, 23]. Analytically, for each partial layer, the matrix operator is written as follows

$$\frac{d\|W\|}{dz} = \|W(z)\|\|K(z)\|, \tag{1.1}$$

where $\|W(z)\|$ is the partial polarization matrix in the plane z and $\|K(z)\|$—differential optical anisotropy matrix.

For a situation of single scattering, when the phenomenon of volume depolarization is absent, and the mechanisms of optical (phase and amplitude) anisotropy are realized in pure form, the differential matrix $\|K(z)\|$ is characterized by the following symmetry [24–27]

$$\|K\| = \begin{Vmatrix} 0 & D_{0,90} & D_{45,135} & D_{\otimes,\oplus} \\ D_{0,90} & 0 & F_{\otimes,\oplus} & -F_{45,135} \\ D_{45,135} & -F_{\otimes,\oplus} & 0 & F_{0,90} \\ D_{\otimes,\oplus} & F_{45,135} & -F_{0,90} & 0 \end{Vmatrix}. \tag{1.2}$$

Here, various partial matrix elements characterize the polarization manifestations of phase shifts between the orthogonal components ($0°$–$90°$; $45°$–$135°$; right—($\otimes$) and left—($\oplus$) circularly polarized components) of the laser radiation amplitude, which are formed by linear and circular birefringence and dichroism:

- $F_{0,90}$, $F_{45,135}$ and $F_{\otimes,\oplus}$—linear and circular birefringence (соотношения (1.3)–(1.5)));
- $D_{0,90}$, $D_{45,135}$ and $D_{\otimes,\oplus}$—linear and circular dichroism (the ratio (1.6)–(1.8))

$$F_{0,90} \equiv \delta_{0,90} = \tfrac{2\pi}{\lambda} \Delta n_{0,90} l; \quad \Delta n_{0,90} = n_0 - n_{90}; \tag{1.3}$$

$$F_{45;135} \equiv \delta_{45;135} = \tfrac{2\pi}{\lambda} \Delta n_{45;135} l; \quad \Delta n_{45;135} = n_{45} - n_{135}; \tag{1.4}$$

$$F_{\otimes,\oplus} \equiv \varphi_{\otimes,\oplus} = \frac{2\pi}{\lambda} \Delta n_{\otimes,\oplus} l; \quad \Delta n_{\otimes,\oplus} = n_\otimes - n_\oplus; \tag{1.5}$$

$$D_{0,90} \equiv \tau_{0;90} = \tfrac{2\pi}{\lambda} \Delta \tau_{0;90} l; \quad \Delta \tau_{0,90} = \tau_0 - \tau_{90}; \tag{1.6}$$

$$D_{45;135} \equiv \tau_{45;135} = \tfrac{2\pi}{\lambda} \Delta \tau_{45;135} l; \quad \Delta \tau_{45;135} = \tau_{45} - \tau_{135}; \tag{1.7}$$

$$D_{\otimes;\oplus} \equiv \chi_{\otimes;\oplus} = \tfrac{2\pi}{\lambda} \Delta \tau_{\otimes;\oplus} l; \quad \Delta \tau_{\otimes;\oplus} = \tau_\otimes - \tau_\oplus; \tag{1.8}$$

Here $\delta_{0;90}$ and $\delta_{45;135}$ are the phase shifts of structural anisotropy the value $\Delta n_{0;90}$ and $\Delta n_{45;135}$; $\varphi_{\otimes;\oplus}$ is the phase shift of optical activity the value $\Delta n_{\otimes;\oplus}$; $\tau_{0;90}$ and $\tau_{45;135}$ are the phase characteristics of anisotropic absorption of orthogonal components amplitudes of laser wave with linear polarization the value $\Delta \tau_{0;90}$ and $\Delta \tau_{45;135}$; $\chi_{\otimes;\oplus}$ are the phase characteristics of the anisotropic absorption of the orthogonal components of the amplitude of the laser wave with circular polarization the value $\Delta \tau_{\otimes;\oplus}$; n_j and τ_j are the refraction and absorption indices; l is the geometrical thickness; λ is the laser beam wavelength.

Further analytical consideration will be carried out using the generalized parameters of linear birefringence (F) and dichroism (D) [28, 48, 49]

$$F = \sqrt{F_{0;90}^2 + F_{45;135}^2}; \tag{1.9}$$

$$D = \sqrt{D_{0;90}^2 + D_{45;135}^2}. \tag{1.10}$$

For a partially depolarizing layer, the partial matrix $\langle \|K(z)\| \rangle$ in expression (1.1) is rewritten in the form of a superposition:

$$\|K(z)\| = \langle \|K(z)\| \rangle + \|\tilde{K}(z)\|. \tag{1.11}$$

Here $\langle \|K(z)\| \rangle$—polarization part, which describes the average values of the elements.

$\|K(z)\|$, $\|\tilde{K}(z)\|$—depolarized part that describes component fluctuations $\|K(z)\|$ [48, 49]. In addition, there is always a relationship between the differential component $\|K(z)\|$ and the Mueller matrix itself $\|W(z)\|$

$$\|W(z)\| = \exp(\|K(z)\|). \tag{1.12}$$

On this basis (1.12) using relations (1.1), (1.2), (1.11), we can obtain a basic expression for determining the expression $\Lambda n(z)$ for determining the polarized $Pl(z)$ and depolarized $Hd(z)$ logarithmic components [29, 30]

$$\Lambda n(z) = \ln\{\|W(z)\|\} = Pl(z) + Hd(z), \tag{1.13}$$

$$Pl(z) = \langle\|K\|\rangle z; \quad Hd(z) = 0,5\|\tilde{K}\|z^2, \tag{1.14}$$

where

$$Pl = 0.5\left(\Lambda n - Mn\Lambda n^T Mn\right); \quad Hd = 0.5\left(\Lambda n + Mn\Lambda n^T Mn\right);$$
$$Mn = diag(1, -1, -1, -1). \tag{1.15}$$

The theoretical analysis (1.2), (1.13)–(1.15) provided an analytical expression for the polarization component of the differential matrix:

$$\langle\|K\|\rangle = z^{-1} \begin{Vmatrix} 0 & (j_{12}+j_{21}) & (j_{13}+j_{31}) & (j_{14}+j_{41}) \\ (j_{21}+j_{12}) & 0 & (j_{23}-j_{32}) & (j_{24}-j_{42}) \\ (j_{31}+j_{13}) & (j_{32}-j_{23}) & 0 & (j_{34}-j_{43}) \\ (j_{41}+j_{14}) & (j_{42}-j_{24}) & (j_{43}-j_{34}) & 0 \end{Vmatrix}, \tag{1.16}$$

where

$$\begin{cases} j_{ik} = \ln W_{ik}; \\ j_{ik}+j_{ki} = \ln(W_{ik} \times W_{ki}); \\ j_{ik}-j_{ki} = \ln\left(\frac{W_{ik}}{W_{ki}}\right) \end{cases} \tag{1.17}$$

In expanded form (taking into account (1.15) and (1.17)), the set of elements of the differential matrix operator $\langle\|K\|\rangle$ (1.16) is written in the following form:

$$\langle\|K\|\rangle = z^{-1} \begin{cases} \langle k_{12}\rangle = \langle k_{21}\rangle = \ln(W_{12}W_{21}); \\ \langle k_{13}\rangle = \langle k_{31}\rangle = \ln(W_{13}W_{31}); \\ \langle k_{14}\rangle = \langle k_{41}\rangle = \ln(W_{14}W_{41}); \\ \langle k_{23}\rangle = -\langle k_{32}\rangle = \ln\left(\frac{W_{23}}{W_{32}}\right); \\ \langle k_{24}\rangle = -\langle k_{42}\rangle = \ln\left(\frac{W_{24}}{W_{42}}\right); \\ \langle k_{34}\rangle = -\langle k_{43}\rangle = \ln\left(\frac{W_{34}}{W_{43}}\right). \end{cases} \tag{1.18}$$

The obtained relations (1.3)–(1.8) and (1.18) are the basis for determining the Mueller-matrix differential algorithm for the polarization reconstruction of layer-by-layer coordinate distributions of the average values of the phase $(\delta; \varphi)$ and amplitude $(\tau; \chi)$ anisotropy parameters with a thickness interval Δz $(0 \leq z \leq l)$ of a biological tissue sample.

$$\delta_{0,90} = \frac{2\pi z}{\lambda} \Delta n_{0,90} = \ln\left(\frac{W_{24}}{W_{42}}\right); \tag{1.19}$$

$$\delta_{45,135} = \frac{2\pi z}{\lambda} \Delta n_{45,135} = \ln\left(\frac{W_{34}}{W_{43}}\right); \tag{1.20}$$

$$\varphi = \frac{2\pi z}{\lambda} \Delta n_{\otimes,\oplus} = \ln\left(\frac{W_{23}}{W_{32}}\right); \tag{1.21}$$

$$\tau_{0,90} = \frac{2\pi z}{\lambda} \Delta \tau_{0;90} = \ln(W_{12} W_{21}); \tag{1.22}$$

$$\tau_{45,135} = \frac{2\pi z}{\lambda} \Delta \tau_{45;135} = \ln(W_{13} W_{31}); \tag{1.23}$$

$$\chi = \frac{2\pi z}{\lambda} \Delta \tau_{\otimes,\oplus} = \ln(W_{14} W_{41}). \tag{1.24}$$

Thus, the obtained algorithm (1.19)–(1.24) makes it possible to experimentally study the average parameters of the optical anisotropy of a partially depolarizing layer of biological tissue. In turn, the information obtained provides the ability to detect changes in the morphological structure (fibrillar networks—linear birefringence and dichroism) and optically active molecular complexes (spiral-like structure of protein molecules—circular birefringence and dichroism).

In parallel, a different mechanism of interaction of the laser probe is realized in the volume of the optically anisotropic layer, which is associated with fluctuations in the magnitude of linear and circular birefringence and dichroism.

Analytically, the scenario of such an interaction describes the depolarized component of the matrix logarithm $\mathrm{Hd}(\tilde{K}_{ik})$. Using ratio (1.13), it is possible to determine the structure of a second-order differential matrix [30–34]. Elements of this operator characterize fluctuations in the parameters of phase and amplitude anisotropy in the volume of the depolarizing biological layer:

$$\|\tilde{K}\| = 0.5 z^{-2} \left\| \begin{array}{cccc} j_{11} & (j_{12} - j_{21}) & (j_{13} - j_{31}) & (j_{14} - j_{41}) \\ (j_{21} - j_{12}) & j_{22} & (j_{23} + j_{32}) & (j_{24} + j_{42}) \\ (j_{31} - j_{13}) & (j_{32} + j_{23}) & j_{33} & (j_{34} + j_{43}) \\ (j_{41} - j_{14}) & (j_{42} + j_{24}) & (j_{43} + j_{34}) & j_{44} \end{array} \right\|. \tag{1.25}$$

In the expanded Mueller-matrix form (1.25) is written as follows

$$\|\tilde{K}\| = 0.5z^{-2} \left\| \begin{matrix} \ln W_{11} & \ln\left(\frac{W_{12}}{W_{21}}\right) & \ln\left(\frac{W_{13}}{W_{31}}\right) & \ln\left(\frac{W_{14}}{W_{41}}\right) \\ \ln\left(\frac{W_{21}}{W_{12}}\right) & \ln W_{22} & \ln(W_{23}W_{32}) & \ln(W_{24}W_{42}) \\ \ln\left(\frac{W_{31}}{W_{13}}\right) & \ln(W_{32}W_{23}) & \ln W_{33} & \ln(W_{34}W_{43}) \\ \ln\left(\frac{W_{41}}{W_{14}}\right) & \ln(W_{42}W_{24}) & \ln(W_{43}W_{34}) & \ln W_{44} \end{matrix} \right\|. \quad (1.26)$$

Further analysis of the obtained (1.26) depolarization operator (2nd order differential matrix) was carried out using V. Devlaminck's theory [50]. Here, to describe (ratio (1.2)–(1.8)) of the six elements (hereinafter polarization properties $\xi_{i=1-6}$) of the differential matrix, a new statistical form is introduced—a superposition of average μ_i and the fluctuating σ_i components

$$\xi_i = \mu_i + \sigma_i. \quad (1.27)$$

Here

$$\mu_i = \begin{pmatrix} \mu_1 = \langle D_{0;90} \rangle; \\ \mu_2 = \langle D_{45;135} \rangle; \\ \mu_3 = \langle D_{\otimes;\oplus} \rangle; \\ \mu_4 = \langle F_{0;90} \rangle; \\ \mu_5 = \langle F_{45;135} \rangle; \\ \mu_6 = \langle F_{\otimes;\oplus} \rangle; \end{pmatrix}; \sigma_i = \begin{pmatrix} \sigma_1 = \sqrt{\theta_1\left(D_{0;90}\right)}; \\ \sigma_2 = \sqrt{\theta_2\left(D_{45;135}\right)}; \\ \sigma_3 = \sqrt{\theta_3\left(D_{\otimes;\oplus}\right)}; \\ \sigma_4 = \sqrt{\theta_4\left(F_{0;90}\right)}; \\ \sigma_5 = \sqrt{\theta_5\left(F_{45;135}\right)}; \\ \sigma_6 = \sqrt{\theta_6\left(F_{\otimes;\oplus}\right)}. \end{pmatrix}, \quad (1.28)$$

where $\sqrt{\theta_i}$—standard deviation of magnitude fluctuations of набора polarization properties $\xi_{i=1-6}$.

In view of (1.27), (1.28), the depolarization component or differential matrix of the 2nd order takes the following form

$$\|\tilde{K}\| = \left\| \begin{matrix} (\sigma_4^2 + \sigma_5^2 + \sigma_6^2)_{11} & -0.5(\sigma_2\sigma_6 - \sigma_3\sigma_5)_{12} & -0.5(\sigma_3\sigma_4 - \sigma_1\sigma_6)_{13} & -0.5(\sigma_1\sigma_5 - \sigma_2\sigma_4)_{14} \\ 0.5(\sigma_2\sigma_6 - \sigma_3\sigma_5)_{21} & (\sigma_4^2 - \sigma_2^2 - \sigma_3^2)_{22} & 0.5(\sigma_1\sigma_2 + \sigma_4\sigma_5)_{23} & 0.5(\sigma_1\sigma_3 + \sigma_4\sigma_5)_{24} \\ 0.5(\sigma_3\sigma_4 - \sigma_1\sigma_6)_{31} & 0.5(\sigma_1\sigma_2 + \sigma_4\sigma_5)_{32} & (\sigma_5^2 - \sigma_1^2 - \sigma_3^2)_{33} & 0.5(\sigma_2\sigma_3 + \sigma_5\sigma_6)_{34} \\ 0.5(\sigma_1\sigma_5 - \sigma_2\sigma_4)_{41} & 0.5(\sigma_1\sigma_3 + \sigma_4\sigma_6)_{42} & 0.5(\sigma_2\sigma_3 + \sigma_5\sigma_6)_{43} & (\sigma_6^2 - \sigma_1^2 - \sigma_3^2)_{44} \end{matrix} \right\|. \quad (1.29)$$

In other words, the differential matrix operator (1.29) is a mathematical analogue of a partially depolarizing phase-inhomogeneous optically anisotropic biological layer with fluctuations of the linear and circular birefringence and dichroism parameters.

Devlamink [50] performed a physical analysis of the partial elements of a second-order differential matrix of such an optically anisotropic object:

- the diagonal matrix elements $\tilde{k}_{11;22;33;44}$ characterize the superposition of the values of the set of statistical moments of the second order (dispersion $\theta_{i=1\div6}$), which characterize the magnitude of the fluctuations of the parameters of phase ($\theta_{4\div6}$) and amplitude ($\theta_{1\div3}$) anisotropy;
- off-diagonal elements $\tilde{k}_{ik;i\neq k}$ are correlated in nature and determine the degree of mutual correlation of the fluctuations of the polarization manifestations of linear and circular birefringence ($\sigma_{4\div6}$) and dichroism ($\sigma_{1\div3}$) of the polycrystalline component of the biological layer.

It was established that an increase in the degree of depolarization (scattering frequency) [51] is accompanied by an increase in the magnitude of fluctuations and a decrease in the degree of mutual correlation of the mechanisms of phase and amplitude anisotropy. As a result of such a physical scenario, the symmetry of the matrix operator (1.29) tends to a diagonal differential matrix $diag\left(\tilde{k}_{11};\ \tilde{k}_{22};\ \tilde{k}_{33};\tilde{k}_{44}\right)$. Based on this, the statistical analysis of 2D maps (tomograms) of the following partial matrix elements is relevant

$$\tilde{T}(\tilde{K}_{ii}) = 0.5z^{-2}\begin{cases} \tilde{k}_{11} = (\theta_4 + \theta_5 + \theta_6) = \ln(W_{11}); \\ \tilde{k}_{22} = (\theta_4 - \theta_2 - \theta_3) = \ln(W_{22}); \\ \tilde{k}_{33} = (\theta_5 - \theta_1 - \theta_3) = \ln(W_{33}); \\ \tilde{k}_{44} = (\theta_6 - \theta_1 - \theta_2) = \ln(W_{44}). \end{cases} \qquad (1.30)$$

Thus, the theoretical differential Mueller-matrix description of the processes of interaction of laser radiation with phase-inhomogeneous depolarizing biological layers allows us to formulate the following principles of differential Mueller-matrix tomography of distributions of average values (polarization-phase tomography) and fluctuation values (diffuse tomography) of linear and circular birefringence and dichroism:

- Traditional vector-parametric mapping of the distributions of the magnitude of the parameters of the Stokes vector images of the studied biological layer and definition of a series of Mueller-matrix images $W_{ik}(x, y)$ [17];
- Coordinate reconstruction by algorithms (relations (1.18) and (1.19)–(1.24)) of the polarization-phase tomography of distribution maps of the average values of

linear and circular birefringence and dichroism $T(\langle k_{ik}\rangle) = \begin{Bmatrix} \delta \\ \varphi \\ \tau \\ \chi \end{Bmatrix}(x, y);$

- Coordinate restoration by diffuse tomography algorithms (relations (1.26) and (1.27)–(1.30)) of the distribution of fluctuations of the parameters of the phase and amplitude anisotropy of the biological layer $\tilde{T}(\tilde{k}_{ii}) = (\theta_4 + \theta_5 + \theta_6)^{-1}$

$$\left.\begin{cases} 1 \\ (\theta_4 - \theta_2 - \theta_3); \\ (\theta_5 - \theta_1 - \theta_3); \\ (\theta_6 - \theta_1 - \theta_2) \end{cases}\right\}(x, y);$$

- Determining the set of magnitude and range of changes in the set of statistical moments of the 1st–4th order [18–22], which characterize the polarization-phase $T(x, y)$ and diffuse $\tilde{T}(x, y)$ tomograms of the depolarizing layer;
- Determination of operational characteristics (sensitivity, specificity, and accuracy) [52–54], which characterize the diagnostic power of polarization-phase and diffuse tomography methods within representative samples of biological tissue samples.

Thus, the application of methods and systems of differential Mueller-matrix mapping of the polycrystalline structure of biological tissue samples extends the functionality of traditional "single-point" laser polarimetry techniques to the most common case of depolarizing or multiple scattering layers.

On the other hand, the task of developing polarization-correlation estimation of the topographic structure of the obtained polarization-phase and diffuse tomograms of the average values and fluctuations of linear and circular birefringence and dichroism of samples of depolarizing biological tissues of different morphological structure and physiological state remains relevant.

1.4 Theory of the Method of Polarization-Correlation Stokes-Polarimetry of the Optical Anisotropy

This part contains the theoretical justification of the Stokes-parametric correlation approach [46, 47] (hereinafter referred to as "Stokes-correlometry") to the analysis of scenarios of the formation of polarization-inhomogeneous object fields of complex amplitudes by depolarizing optically anisotropic biological structures:

1. Traditionally, in the description of correlation relationships between spatially separated points of the field of complex amplitudes, the matrix operator is used —mutual spectral density matrix:

$$W_{i,j}(r_1, r_2) = E_i^*(r_1) \cdot E_j(r_2), i, j = x, y. \tag{1.31}$$

Here, r_1 and r_2 are the coordinates of the spatially separated points in the polarization-inhomogeneous vector field.

Using (1.31), we obtain a set of analytical expressions for the "two-point" Stokes vector parameters (hereinafter the Stokes-correlometry parameters—SCP):

$$Sv_1 = V_{xx}(r_1, r_2) + V_{yy}(r_1, r_2), \tag{1.32}$$

$$Sv_2 = V_{xx}(r_1, r_2) - V_{yy}(r_1, r_2), \tag{1.33}$$

$$Sv_3 = V_{xy}(r_1, r_2) - V_{yx}(r_1, r_2), \tag{1.34}$$

$$Sv_4 = i\left[V_{yx}(r_1, r_2) + V_{xy}(r_1, r_2)\right]. \tag{1.35}$$

where

$$\begin{cases} V_{xx}(r_1, r_2) = E_x^*(r_1)E_x(r_2); \\ V_{yy}(r_1, r_2) = E_y^*(r_1)E_y(r_2); \\ V_{xy}(r_1, r_2) = E_x^*(r_1)E_y(r_2); \\ V_{yx}(r_1, r_2) = E_y^*(r_1)E_x(r_2). \end{cases} \tag{1.36}$$

In the expanded form, relations (1.32)–(1.36), which describe the set of SCPs, take the following form:

$$Sv_1(r_1, r_2) = E_x^*(r_1)E_x(r_2) + E_y^*(r_1)E_y(r_2); \tag{1.37}$$

$$Sv_2(r_1, r_2) = E_x^*(r_1)E_x(r_2) - E_y^*(r_1)E_y(r_2); \tag{1.38}$$

$$Sv_3(r_1, r_2) = E_x^*(r_1)E_y(r_2) + E_y^*(r_1)E_x(r_2); \tag{1.39}$$

$$Sv_4(r_1, r_2) = i\left[E_y^*(r_1)E_x(r_2) - E_x^*(r_1)E_y(r_2)\right]. \tag{1.40}$$

2. The next step of the analytical review is to determine the relationship between the parameters of optical anisotropy and the value of complex amplitudes $E(r)$ of each point r. Using the Jones matrix formalism, we can obtain a relation that characterizes this relationship:

$$E(r) = \begin{pmatrix} E_x \\ E_y \end{pmatrix}(r) = \begin{pmatrix} |E_x| \\ |E_y|\exp(\varphi_y - \varphi_x) \end{pmatrix}(r)$$
$$= \begin{pmatrix} 1 \\ tg\gamma\exp(\varphi) \end{pmatrix}(r) = \begin{pmatrix} 1 \\ tg\gamma(\cos\varphi + i\sin\varphi) \end{pmatrix}(r) \tag{1.41}$$

Here, $tg\gamma(r) = \frac{|E_y|(r)}{|E_x|(r)}$ and $\varphi(r) = (\varphi_y - \varphi_x)(r)$—phase shifts between the linearly polarized orthogonal components $(|E_x|(r), |E_y|(r))$ amplitude at the point r of the object field of the biological layer.

3. The resulting relationship (1.41) makes it possible to determine the expressions of the SCP set (Eqs. (1.37)–(1.40)) in the form of functionals of the phase anisotropy parameters:

$$Sv_1(r_1, r_2) = (1 + tg\gamma_1 tg\gamma_2 \cos(\varphi_1 - \varphi_2)) - i(tg\gamma_1 tg\gamma_2 \sin(\varphi_1 - \varphi_2)); \quad (1.42)$$

$$Sv_2(r_1, r_2) = (1 - tg\gamma_1 tg\gamma_2 \cos(\varphi_1 - \varphi_2)) + i(tg\gamma_1 tg\gamma_2 \sin(\varphi_1 - \varphi)); \quad (1.43)$$

$$Sv_3(r_1, r_2) = (tg\gamma_2 \cos\varphi_2 + tg\gamma_1 \cos\varphi_1) + i(tg\gamma_2 \sin\varphi_2 - tg\gamma_1 \sin\varphi_1); \quad (1.44)$$

$$Sv_4(r_1, r_2) = (tg\gamma_1 \sin\varphi_1 + tg\gamma_2 \sin\varphi_2) + i(tg\gamma_1 \cos\varphi_1 - tg\gamma_2 \cos\varphi_2). \quad (1.45)$$

4. The resulting complex expressions of SCP (1.42)–(1.45) can be characterized by valid parameters–modulus $\left|Sv_{i=1;2;3;4}\right|$ and the phase $ArgSv_{i=1;2;3;4}$:

$$\begin{cases} |Sv_1|(r_1, r_2) = \left((1 + tg\gamma_1 tg\gamma_2 \cos(\varphi_1 - \varphi_2))^2 + (tg\gamma_1 tg\gamma_2 \sin(\varphi_1 - \varphi_2))^2\right)^{0.5}; \\ ArgSv_2(r_1, r_2) = arctg\left(\frac{tg\gamma_1 tg\gamma_2 \sin(\varphi_1 - \varphi_2)}{(1 + tg\gamma_1 tg\gamma_2 \cos(\varphi_1 - \varphi_2))}\right); \end{cases}$$

$$(1.46)$$

$$\begin{cases} |Sv_2|(r_1, r_2) = \left((1 - tg\gamma_1 tg\gamma_2 \cos(\varphi_1 - \varphi_2))^2 + (tg\gamma_1 tg\gamma_2 \sin(\varphi_1 - \varphi_2))^2\right)^{0.5}; \\ ArgSv_2(r_1, r_2) = arctg\left(\frac{tg\gamma_1 tg\gamma_2 \sin(\varphi_1 - \varphi_2)}{(1 - tg\gamma_1 tg\gamma_2 \cos(\varphi_1 - \varphi_2))}\right); \end{cases}$$

$$(1.47)$$

$$\begin{cases} |Sv_3|(r_1, r_2) = \left((tg\gamma_2 \cos\varphi_2 + tg\gamma_1 \cos\varphi_1)^2 + (tg\gamma_2 \sin\varphi_2 - tg\gamma_1 \sin\varphi_1)^2\right)^{0.5}; \\ ArgSv_3(r_1, r_2) = arctg\left(\frac{(tg\gamma_2 \sin\varphi_2 - tg\gamma_1 \sin\varphi_1)}{(tg\gamma_2 \cos\varphi_2 + tg\gamma_1 \cos\varphi_1)}\right); \end{cases}$$

$$(1.48)$$

$$\begin{cases} |Sv_4|(r_1, r_2) = \left((tg\gamma_1 \sin\varphi_1 + tg\gamma_2 \sin\varphi_2)^2 + (tg\gamma_1 \cos\varphi_1 - tg\gamma_2 \cos\varphi_2)^2\right)^{0.5}; \\ ArgSv_4(r_1, r_2) = arctg\left(\frac{(tg\gamma_1 \cos\varphi_1 - tg\gamma_2 \cos\varphi_2)}{(tg\gamma_1 \sin\varphi_1 + tg\gamma_2 \sin\varphi_2)}\right). \end{cases}$$

$$(1.49)$$

5. Of greatest practical interest to us is the situation of weak phase fluctuations ($\varphi_i \leq 0.12$ rad; $\cos\varphi_i \to 1$; $\sin\varphi_i \to \varphi_i$) of an optically anisotropic biological layer. To this end, we rewrite relations (1.46)–(1.49) in this approximation:

$$\begin{cases} |Sv_1|(r_1, r_2) = \left((1 + tg\gamma_1 tg\gamma_2)^2 + (tg\gamma_1 tg\gamma_2(\varphi_1 - \varphi_2))^2\right)^{0.5}; \\ ArgSv_2(r_1, r_2) = arctg\left(\frac{tg\gamma_1 tg\gamma_2(\varphi_1 - \varphi_2)}{(1 + tg\gamma_1 tg\gamma_2)}\right); \end{cases} \quad (1.50)$$

$$\begin{cases} |Sv_2|(r_1, r_2) = \left((1 - tg\gamma_1 tg\gamma_2)^2 + (tg\gamma_1 tg\gamma_2(\varphi_1 - \varphi_2))^2\right)^{0.5}; \\ ArgSv_2(r_1, r_2) = arctg\left(\frac{(tg\gamma_1 tg\gamma_2(\varphi_1 - \varphi_2))}{(1 - tg\gamma_1 tg\gamma_2)}\right); \end{cases} \quad (1.51)$$

$$\begin{cases} |Sv_3|(r_1, r_2) = \left((tg\gamma_2 + tg\gamma_1)^2 + (\varphi_2 tg\gamma_2 - \varphi_1 tg\gamma_1)^2\right)^{0.5}; \\ ArgSv_3(r_1, r_2) = arctg\left(\frac{(\varphi_2 tg\gamma_2 - \varphi_1 tg\gamma_1)}{(tg\gamma_2 + tg\gamma_1)}\right); \end{cases} \tag{1.52}$$

$$\begin{cases} |Sv_4|(r_1, r_2) = \left((\varphi_2 tg\gamma_2 + \varphi_1 tg\gamma_1)^2 + (tg\gamma_1 + tg\gamma_2)^2\right)^{0.5}; \\ ArgSv_4(r_1, r_2) = arctg\left(\frac{(tg\gamma_1 + tg\gamma_2)}{(\varphi_2 tg\gamma_2 + \varphi_1 tg\gamma_1)}\right). \end{cases} \tag{1.53}$$

The found Stokes correlometry algorithms (relations (1.50)–(1.53)) show:

i. the value of the object field modulus $|Sv_{1-4}|(r_1, r_2)$ of a weakly anisotropic biological medium is predominantly determined by the coordination of the directions of the optical axes $\rho_{1;2}$ of biological crystallites;

ii. the value of the phase $ArgSv_{1-4}(r_1, r_2)$ of the set of SCP spatially separated points of the polarization-inhomogeneous field is determined by the superposition of phase shifts $\varphi_{1;2}$ and directions of the optical axes $\gamma_{1;2}$ of various points of the polycrystalline component of the biological layer.

6. In the future, our consideration of the methods of polarization Stokes correlometry (without reducing the completeness of the analysis) will be based on the transformation of the 3rd and the 4th "two-point" Stokes vector parameters. This choice is due to the fact that the "one-point" the third $Sv_3(r)$ and fourth Stokes-vector parameters $Sv_4(r)$ are responsible for the characterization of the formation of a polarization-inhomogeneous field (azimuth distribution and polarization ellipticity) of an optically anisotropic biological preparation.

7. In order to build the principles of physical analysis of Stokes correlometry data, the characteristic values and ranges of the values of the third and the fourth parameters of the two-point Stokes vector were determined.

Table 1.1 presents the characteristic values of the "two-point" parameters $Sv_{3;4}(r_1, r_2)$ (relations (1.7)–(1.15)) and module $|Sv_{3;4}|(r_1, r_2)$ and phase $ArgSv_{3;4}(r_1, r_2)v$ (relations (1.17)–(1.19)) for different polarization states at the spatially separated points r_1, r_2 of the polarized an inhomogeneous object field of an optically anisotropic biological layer.

The data given in Table 1.1 determines the range of variation of the module $|Sv_{3;4}|(r_1, r_2)$ and phase $ArgSv_{3;4}(r_1, r_2)$ depending on the correlation of the directions of the optical axes and phase shifts $(\gamma_{1;2}, \varphi_{1;2})$ at different points (r_1, r_2) of the polycrystalline layer of a biological sample:

$$0 \leq |Sv_{3;4}|(r_1, r_2) \leq 1; \tag{1.54}$$

$$0 \leq ArgSv_{3;4}(r_1, r_2) \leq 0.5\pi. \tag{1.55}$$

Table 1.1 Characteristic values of the modulus $|Sv_{3;4}|(r_1, r_2)$ and phase $ArgSv_{3;4}(r_1, r_2)$

γ_1, rad; φ_1, rad	$E(r_1)$	γ_2, rad; φ_2, rad	$E(r_2)$	$|Sv_3|(r_1, r_2)$	$ArgSv_3(r_1, r_2)$	$|Sv_4|(r_1, r_2)$	$ArgSv_4(r_1, r_2)$
"Collinear" states of polarization in points r_1, r_2							
$\gamma_1 = 0;$ $\varphi_1 = 0$	$\begin{pmatrix} 1 \\ 0 \end{pmatrix}$	$\gamma_2 = 0;$ $\varphi_2 = 0$	$\begin{pmatrix} 1 \\ 0 \end{pmatrix}$	0	"-"	0	"-"
$\gamma_1 = 0.25\pi;$ $\varphi_1 = 0$	$\frac{1}{\sqrt{2}}\begin{pmatrix} 1 \\ 1 \end{pmatrix}$	$\gamma_2 = 0.25\pi;$ $\varphi_2 = 0$	$\frac{1}{\sqrt{2}}\begin{pmatrix} 1 \\ 1 \end{pmatrix}$	1	0	0	"-"
$\gamma_1 = 0.25\pi;$ $\varphi_1 = 0.5\pi$	$\frac{1}{\sqrt{2}}\begin{pmatrix} 1 \\ i \end{pmatrix}$	$\gamma_2 = 0.25\pi;$ $\varphi_2 = 0.5\pi$	$\frac{1}{\sqrt{2}}\begin{pmatrix} 1 \\ i \end{pmatrix}$	1	0	1	0
"Intermediate" states of polarization in points r_1, r_2							
$\gamma_1 = 0;$ $\varphi_1 = 0$	$\begin{pmatrix} 1 \\ 0 \end{pmatrix}$	$\gamma_2 = 0.25\pi;$ $\varphi_2 = 0$	$\frac{1}{\sqrt{2}}\begin{pmatrix} 1 \\ 1 \end{pmatrix}$	$\frac{1}{\sqrt{2}}$	0	$\frac{1}{\sqrt{2}}$	0.5π
$\gamma_1 = 0.25\pi;$ $\varphi_1 = 0$	$\begin{pmatrix} 1 \\ 0 \end{pmatrix}$	$\gamma_2 = 0.25\pi;$ $\varphi_2 = 0$	$\frac{1}{\sqrt{2}}\begin{pmatrix} 1 \\ 1 \end{pmatrix}$	$\frac{1}{\sqrt{2}}$	0	$\frac{1}{\sqrt{2}}$	0.5π
$\gamma_1 = 0.25\pi;$ $\varphi_1 = 0.5\pi$	$\begin{pmatrix} 1 \\ 0 \end{pmatrix}$	$\gamma_2 = 0.25\pi;$ $\varphi_2 = 0.5\pi$	$\frac{1}{\sqrt{2}}\begin{pmatrix} 1 \\ 1 \end{pmatrix}$	$\frac{1}{\sqrt{2}}$	0	$\frac{1}{\sqrt{2}}$	0.5π
"Orthogonal" states of polarization in points r_1, r_2							
$\gamma_1 = 0;$ $\varphi_1 = 0$	$\begin{pmatrix} 1 \\ 0 \end{pmatrix}$	$\gamma_2 = 0.25\pi;$ $\varphi_2 = 0$	$\begin{pmatrix} 0 \\ 1 \end{pmatrix}$	1	0	1	0.5π
$\gamma_1 = 0.25\pi;$ $\varphi_1 = 0$	$\frac{1}{\sqrt{2}}\begin{pmatrix} 1 \\ 1 \end{pmatrix}$	$\gamma_2 = -0.25\pi;$ $\varphi_2 = 0$	$\frac{1}{\sqrt{2}}\begin{pmatrix} 1 \\ -1 \end{pmatrix}$	0	"-"	1	0.5π
$\gamma_1 = 0.25\pi;$ $\varphi_1 = 0.5\pi$	$\frac{1}{\sqrt{2}}\begin{pmatrix} 1 \\ i \end{pmatrix}$	$\gamma_2 = 0.25\pi;$ $\varphi_2 = -0.5\pi$	$\frac{1}{\sqrt{2}}\begin{pmatrix} 1 \\ -i \end{pmatrix}$	0	"-"	0	"-"

Subsequently, to analyze the experimental data of Stokes correlometry of polarization-inhomogeneous object fields of preparations of real biological tissues, the coordinate distributions of the magnitude of the module $\left|Sv_{i=3;4}(\Delta x; \Delta y)\right|$ will be called SCP "orientation correlation maps", and the phases $Arg\left(Sv_{i=3;4}(\Delta x; \Delta y)\right)$—the SCP "phase correlation maps".

All the experimental and analytical results obtained by measuring and analyzing the Mueller matrices are represented in 2D ("Mueller-matrix image" [1–13], "Polarization-phase tomogram" [17–19], "Diffuse tomogram" [48, 49], "Orientation correlation map" [14–16, 55–58] and "Phase correlation map" [41, 44, 45]) formats. Therefore, it is important to develop a new technique of complex polarimetric 3D diagnostics of polycrystalline networks of biological tissues. Such an opportunity can provide a combination of traditional methods of Mueller-matrix and holographic mapping of phase-inhomogeneous layers.

1.5 Analytical Foundations of 3D Mueller-Matrix Mapping

The use of a reference wave of laser radiation, which in the scheme of optical interferometer is superimposed on a polarization ally inhomogeneous image of a biological layer is fundamental. The resulting interference pattern is recorded using a digital camera. With the use of Furies transform integrals the operation of digital holographic reproduction of distributions of complex amplitudes $\{E_x(x, y); E_y(x, y)\}$ of the objective field of a biological layer is performed.

For each state of the irradiating beam, the reconstructed distributions of the Stokes vector parameters of the object field of a biological layer are calculated according to the reproduced distributions of complex amplitudes $\{E_x(x, y); E_y(x, y)\}$

$$
\begin{aligned}
Sv_1(0°, 90°, 45°, \otimes) &= |E_x|^2 + |E_y|^2; \\
Sv_2(0°, 90°, 45°, \otimes) &= |E_x|^2 - |E_y|^2; \\
Sv_3(0°, 90°, 45°, \otimes) &= 2\mathrm{Re}\left|E_x E_y^*\right|; \\
Sv_4(0°, 90°, 45°, \otimes) &= 2\mathrm{Im}\left|E_x E_y^*\right|.
\end{aligned}
\tag{1.56}
$$

Here $0°, 90°, 45°$—polarization azimuths of linearly polarized irradiating beams, $\otimes$—the right-circularly polarized beam.

Further, on the basis of relations (1.56), the set of Mueller matrix elements is calculated by the following Stokes-polarimetric relations:

- for Stokes vectors of linearly polarized probing beams $Sv^0(0°); Sv^0(90°)$:

$$\left\{\begin{array}{l} \left[Sv^0(0°) = \{W\}\begin{pmatrix}1\\1\\0\\0\end{pmatrix} \rightarrow Sv(0°) = \begin{pmatrix}W_{11}+W_{12}\\W_{21}+W_{22}\\W_{31}+W_{32}\\W_{41}+W_{42}\end{pmatrix}\right];\\[2em] \left[S^0v(90°) = \{W\}\begin{pmatrix}1\\-1\\0\\0\end{pmatrix} \rightarrow Sv(90°) = \begin{pmatrix}W_{11}-W_{12}\\W_{21}-W_{22}\\W_{31}-W_{32}\\W_{41}-W_{42}\end{pmatrix}\right]\end{array}\right\} \Rightarrow W_{ik}$$

$$= \left\|\begin{array}{cc}W_{11} & W_{12}\\W_{21} & W_{22}\\W_{31} & W_{32}\\W_{41} & W_{42}\end{array}\right\|;$$

$$(1.57)$$

- for Stokes vectors of linearly polarized probing beams $Sv^0(45°); Sv^0(135°)$:

$$\left\{\begin{array}{l} \left[Sv^0(45°) = \{W\}\begin{pmatrix}1\\0\\1\\0\end{pmatrix} \rightarrow Sv(45°) = \begin{pmatrix}W_{11}+W_{13}\\W_{21}+W_{23}\\W_{31}+W_{33}\\W_{41}+W_{43}\end{pmatrix}\right];\\[2em] \left[S^0v(135°) = \{W\}\begin{pmatrix}1\\0\\-1\\0\end{pmatrix} \rightarrow Sv(135°) = \begin{pmatrix}W_{11}-W_{13}\\W_{21}-W_{23}\\W_{31}-W_{33}\\W_{41}-W_{43}\end{pmatrix}\right]\end{array}\right\} \Rightarrow W_{ik}$$

$$= \left\|\begin{array}{cc}W_{11} & W_{13}\\W_{21} & W_{23}\\W_{31} & W_{33}\\W_{41} & W_{43}\end{array}\right\|;$$

$$(1.58)$$

- for Stokes vectors of right- and left-circularly polarized probing beams $Sv^0(\otimes); \; Sv^0(\oplus)$:

$$
\left\{
\begin{aligned}
&\left[Sv^0(\otimes) = \{W\} \begin{pmatrix} 1 \\ 0 \\ 0 \\ 1 \end{pmatrix} \to Sv(\otimes) = \begin{pmatrix} W_{11} + W_{14} \\ W_{21} + W_{24} \\ W_{31} + W_{34} \\ W_{41} + W_{44} \end{pmatrix} \right]; \\
&\left[Sv^0(\oplus) = \{W\} \begin{pmatrix} 1 \\ 0 \\ 0 \\ -1 \end{pmatrix} \to Sv(\oplus) = \begin{pmatrix} W_{11} - W_{14} \\ W_{21} - W_{24} \\ W_{31} - W_{34} \\ W_{41} - W_{44} \end{pmatrix} \right]
\end{aligned}
\right\} \Rightarrow W_{ik}
$$

$$
= \begin{Vmatrix} W_{11} & W_{14} \\ W_{21} & W_{24} \\ W_{31} & W_{34} \\ W_{41} & W_{44} \end{Vmatrix}; \tag{1.59}
$$

Expressions (1.56)–(1.59) result in working relations for determining the values of Mueller matrix elements

$$
\{W\} = 0.5 \begin{Vmatrix}
\left(Sv_1^0 + Sv_1^{90}\right) & \left(Sv_1^0 - Sv_1^{90}\right) & \left(Sv_1^{45} - Sv_1^{135}\right) & \left(Sv_1^{\otimes} - Sv_1^{\oplus}\right) \\
\left(Sv_2^0 + Sv_2^{90}\right) & \left(Sv_2^0 - Sv_2^{90}\right) & \left(Sv_2^{45} - Sv_2^{135}\right) & \left(Sv_2^{\otimes} - Sv_2^{\oplus}\right) \\
\left(Sv_3^0 + Sv_3^{90}\right) & \left(Sv_3^0 - Sv_3^{90}\right) & \left(Sv_3^{45} - Sv_3^{135}\right) & \left(Sv_3^{\otimes} - Sv_3^{\oplus}\right) \\
\left(Sv_4^0 + Sv_4^{90}\right) & \left(Sv_4^0 - Sv_4^{90}\right) & \left(Sv_4^{45} - Sv_4^{135}\right) & \left(Sv_4^{\otimes} - Sv_4^{\oplus}\right)
\end{Vmatrix}. \tag{1.60}
$$

Traditionally, in the right side of expressions (1.60), the value of the integral across the entire thickness l of the biological layer of phase shift $\delta(z = l)$ "appears"

$$
Sv(0°, 90°, 45°, \otimes) = \begin{pmatrix}
Sv_1(0°, 90°, 45°, \otimes) \\
Sv_2(0°, 90°, 45°, \otimes) \\
Sv_3(0°, 90°, 45°, \otimes) \\
Sv_4(0°, 90°, 45°, \otimes)
\end{pmatrix} = \begin{pmatrix}
|E_x|^2 + |E_y|^2; \\
|E_x|^2 - |E_y|^2; \\
2|E_x||E_y| \cos \delta(z = l); \\
2|E_x||E_y| \sin \delta(z = l).
\end{pmatrix}
$$
$$
= Sv(E_x, E_y, \delta). \tag{1.61}
$$

Therefore, the direct Muller-matrix mapping results in two-dimensional distributions of the values of matrix elements $W_{ik} = q_j\left(Sv_{i=1;2;3;4}(x, y, \delta)\right)$, averaging $(\delta(z = l))$ over the entire thickness l of the biological layer.

In the case of using a coherent reference wave and algorithms of digital holographic reproduction, it is possible to reconstruct the distributions of complex amplitudes $|E_x| \exp i(\Delta\varphi_x);$ $|E_y| \exp i(\Delta\varphi_y)$ of the object field in a discrete

$(\Delta\varphi_{l=0...q})$ set of phase planes $\varphi_j = \begin{pmatrix} 0 \\ \Delta\varphi \\ 2\Delta\varphi \\ . \\ . \\ \delta \end{pmatrix}$. Due to this, one can obtain a set of

layer-by-layer distributions of the values of matrix elements $(x, y, k\Delta\varphi)$ and determine their volumetric structure

$$W_{ik} = \left\{ g_s\left(|E_x|, |E_y|, \varphi^*\right) \right\} \left(x, y, \begin{pmatrix} 0 \\ \Delta\varphi \\ . \\ \varphi \end{pmatrix} \right). \tag{1.62}$$

Thus, the presented original 3D Mueller-matrix mapping technique (relations (1.56)–(1.62)) expands the functionality of the differential Mueller-matrix mapping (relations (1.1)–(1.30)) and polarization correlometry (relations (1.31)–(1.55)) by obtaining a series of layer-by-layer distributions of the magnitude and phase of the "two-point" parameters of the Stokes vector, polarization-phase and diffuse tomograms of depolarizing layers of optically anisotropic biological tissues.

References

1. Tuchin, V., Wang, L., Zimnjakov, D.: Optical polarization in biomedical applications. Springer, New York, USA (2006)
2. Chipman, R.: Polarimetry. In: Bass, M. (eds.) Handbook of Optics: Vol. I—Geometrical and Physical Optics, Polarized Light, Components and Instruments, pp. 22.1–22.37. McGraw-Hill Professional, New York (2010)
3. Ghosh, N., Wood, M., Vitkin, A.: Polarized light assessment of complex turbid media such as biological tissues via Mueller matrix decomposition. In: Tuchin V (ed.) Handbook of Photonics for Biomedical Science, pp. 253–282. CRC Press, Taylor & Francis Group, London (2010)
4. Jacques, S.: Polarized light imaging of biological tissues. In: Boas, D., Pitris, C., Ramanujam, N. (eds.) Handbook of biomedical optics, pp. 649–669. CRC Press, Boca Raton, London, New York (2011)
5. Ghosh, N.: Tissue polarimetry: concepts, challenges, applications, and outlook. J. Biomed. Opt. 16(11), 110801 (2011)
6. Swami, M., Patel, H., Gupta, P.: Conversion of 3×3 Mueller matrix to 4×4 Mueller matrix for non-depolarizing samples. Opt. Commun. 286, 18–22 (2013)
7. Layden, D., Ghosh, N., Vitkin, A.: Quantitative polarimetry for tissue characterization and diagnosis. In: Wang, R., Tuchin, V. (eds.) Advanced Biophotonics: Tissue Optical Sectioning, pp. 73–108. CRC Press, Taylor & Francis Group, Boca Raton, London, New York. (2013)
8. Vo-Dinh, T.: Biomedical Photonics Handbook, 3 vol. set, 2nd edn. CRC Press, Boca Raton (2014)

9. Vitkin, A., Ghosh, N., Martino, A.: Tissue polarimetry. In: Andrews, D. (ed.) Photonics: Scientific Foundations, Technology and Applications, 4th edn, pp. 239–321. Wiley, Hoboken, New Jersey (2015)

10. Tuchin, V.: Tissue optics: light scattering methods and instruments for medical diagnosis, 2nd edn. SPIE Press, Bellingham, Washington, USA (2007)

11. Bickel, W., Bailey, W.: Stokes vectors, Mueller matrices, and polarized scattered light. Am. J. Phys. **53**(5), 468–478 (1985)

12. Doronin, A., Macdonald, C., Meglinski, I.: Propagation of coherent polarized light in turbid highly scattering medium. J. Biomed. Opt. **19**(2), 025005 (2014)

13. Doronin, A., Radosevich, A., Backman, V., Meglinski, I.: Two electric field Monte Carlo models of coherent backscattering of polarized light. J. Opt. Soc. Am. A **31**(11), 2394 (2014)

14. Arun Gopinathan, P., Kokila, G., Jyothi, M., Ananjan, C., Pradeep, L., Humaira Nazir, S.: Study of collagen birefringence in different grades of oral squamous cell carcinoma using picrosirius red and polarized light microscopy. Scientifica 802980 (2015)

15. Rich, L., Whittaker, P.: Collagen and picrosirius red staining: a polarized light assessment of fibrillar hue and spatial distriburuon. Braz. J. Morphol. Sci, (2005). www.patologia.medicina. ufrj.br

16. Bancelin, S., Nazac, A., Ibrahim, B.H., et al.: Determination of collagen fiber orientation in histological slides using Mueller microscopy and validation by second harmonic generation imaging. Opt. Express **22**(19), 22561–22574 (2014)

17. Ushenko, A., Pishak, V.: Laser polarimetry of biological tissue: principles and applications. In: Tuchin, V. (ed.) Handbook of Coherent-Domain Optical Methods: Biomedical Diagnostics, Environmental and Material Science, pp. 93–138 (2004)

18. Angelsky, O., Ushenko, A., Ushenko, Y., Pishak, V., Peresunko, A.: Statistical, correlation and topological approaches in diagnostics of the structure and physiological state of birefringent biological tissues. In: Handbook of Photonics for Biomedical Science, pp. 283–322 (2010)

19. Ushenko, Y., Boychuk, T., Bachynsky, V., Mincer, O.: Diagnostics of structure and physiological state of birefringent biological tissues: statistical, correlation and topological approaches. In: Tuchin, V (ed.) Handbook of Coherent-Domain Optical Methods. Springer (2013)

20. Angelsky, O., Tomka, Y., Ushenko, A., Ushenko, Y., Yermolenko, S.: 2-D tomography of biotissue images in pre-clinic diagnostics of their pre-cancer states. Proc. SPIE **5972**, 158–162 (2005)

21. Angelsky, O., Ushenko, A., Ushenko, Y.: Investigation of the correlation structure of biological tissue polarization images during the diagnostics of their oncological changes. Phys. Med. Biol. **50**(20), 4811–4822 (2005)

22. Ushenko, Y., Ushenko, V., Dubolazov, A., Balanetskaya, V., Zabolotna, N.: Mueller-matrix diagnostics of optical properties of polycrystalline networks of human blood plasma. Opt. Spectrosc. **112**(6), 884–892 (2012)

23. Ushenko, V., Dubolazov, O., Karachevtsev, A.: Two wavelength Mueller matrix reconstruction of blood plasma films polycrystalline structure in diagnostics of breast cancer. Appl. Opt. **53**(10), B128 (2016)

24. Ushenko, Y., Koval, G., Ushenko, A., Dubolazov, O., Ushenko, V., Novakovskaia, O.: Mueller-matrix of laser-induced autofluorescence of polycrystalline films of dried peritoneal fluid in diagnostics of endometriosis. J. Biomed. Opt. **21**(7), 071116 (2016)

25. Ushenko, A., Dubolazov, A., Ushenko, V., Novakovskaya, O.: Statistical analysis of polarization-inhomogeneous Fourier spectra of laser radiation scattered by human skin in the tasks of differentiation of benign and malignant formations. J. Biomed. Opt. **21**(7), 071110 (2016)

26. Prysyazhnyuk, V., Ushenko, Y., Dubolazov, A., Ushenko, A., Ushenko, V.: Polarization-dependent laser autofluorescence of the polycrystalline networks of blood plasma films in the task of liver pathology differentiation. Appl. Opt. **55**(12), B126 (2016)

27. Azzam, R.: Propagation of partially polarized light through anisotropic media with or without depolarization: A differential 4 × 4 matrix calculus. J. Opt. Soc. Am. **68**(12), 1756 (1978)
28. Jones, R.: A new, "Calculus for the Treatment of Optical Systems VII Properties of the N-Matrices". J. Opt. Soc. Am. **38**(8), 671 (1948)
29. Ortega-Quijano, N., Arce-Diego, J.: Mueller matrix differential decomposition. Opt. Lett. **36** (10), 1942–1944 (2011)
30. Devlaminck, V.: Physical model of differential Mueller matrix for depolarizing uniform media. J. Opt. Soc. Am. A **30**(11), 2196 (2013)
31. Ossikovski, R., Devlaminck, V.: General criterion for the physical reliability of the differential Mueller matrix. Opt. Lett. **39**(5), 1216 (2014)
32. Devlaminck, V., Ossikovski, R.: Uniqueness of the differential Mueller matrix of uniform homogeneous media. Opt. Lett. **39**(11), 3149 (2014)
33. Ossikovski, R., Arteaga, O.: Statistical meaning of the differential Mueller matrix of depolarizing homogeneous media. Opt. Lett. **39**(15), 4470 (2014)
34. Ossikovski, R.: Differential matrix formalism for depolarizing anisotropic media. Opt. Lett. **36**(12), 2330 (2011)
35. Ushenko, V., Pavlyukovich, N., Trifonyuk, L.: Spatial-frequency azimuthally stable cartography of biological polycrystalline networks. Int. J. Opt. **2013**, 1–7 (2013)
36. Pérez-Cárceles, M., Noguera, J., Jiménez, J., Martínez, P., Luna, A., Osuna, E.: Diagnostic efficacy of biochemical markers in diagnosis post-mortem of ischaemic heart disease. Forensic Sci. Int. **142**(1), 1–7 (2004)
37. Martínez Díaz, F., Rodríguez-Morlensín, M., Pérez-Cárceles, M., Noguera, J., Luna, A., Osuna, E.: Biochemical analysis and immunohistochemical determination of cardiac troponin for the postmortem diagnosis of myocardial damage. Histol. Histopathol. **20**(2), 475–481 (2005)
38. Göksedef, B.P.Ç., Akbayır, Ö., Çorbacıoğlu, A., Güraslan, H., Şencan, F., Erol, O., Çetin, A.: J. Turk Ger Gynecol. Assoc. **13**(2), 106–110 (2012)
39. Sharon, A.S., Ritu, S.B., Thomas, E.R., et al.: Striae and pelvic relaxation: two disorders of connective tissue with strong association. J. Invest. Dermatol. **126**, 1745–1748 (2006)
40. Ushenko, VA., Gavrylyak, M.S.: Azimuthally invariant Mueller-matrix mapping of biological tissue in differential diagnosis of mechanisms protein molecules networks anisotropy. Proc. SPIE **8812**, 88120Y (2013)
41. Ushenko, V.A., Gorsky, M.P.: Complex degree of mutual anisotropy of linear birefringence and optical activity of biological tissues in diagnostics of prostate cancer. Opt. Spectrosc. **115**, 290–297 (2013)
42. Ushenko, V.A., Dubolazov, A.V.: Correlation and self similarity structure of polycrystalline network biological layers Mueller matrices images. Proc. SPIE **8856**, 88562D (2013)
43. Ushenko, V.A., Pavlyukovich, N.D., Trifonyuk, L.: Spatial-frequency azimuthally stable cartography of biological polycrystalline networks. Int. J. Opt. **683174**, 2013 (2013)
44. Ushenko, V.A.: Complex degree of mutual coherence of biological liquids. Proc. SPIE **8882**, 88820V (2013)
45. Ushenko, Y.A., et al.: Jones-matrix mapping of complex degree of mutual anisotropy of birefringent protein networks during the differentiation of myocardium necrotic changes. Appl. Opt. **55**, B113–B119 (2016)
46. Tervo, J., Setala, T., Friberg, A.: Degree of coherence for electromagnetic fields. Opt. Expr. **11**, 1137–1143 (2003)
47. Tervo, J., Setala, T., Friberg, A.: Two-point Stokes parameters: interpretation and properties. Opt. Lett. **34**, 3074–3076 (2009)
48. Lu, S.Y., Chipman, R.A.: Interpretation of Mueller matrices based on polar decomposition. J. Opt. Soc. Am. A **13**, 1106–1113 (1996)
49. Guo, Y., Zeng, N., He, H., Yun, T., Du, E., Liao, R., Ma, H.: A study on forward scattering Mueller matrix decomposition in anisotropic medium. Opt. Exp. **21**, 18361–18370 (2013)

50. Ushenko, Y.A., et al.: Spatial-frequency Fourier polarimetry of the complex degree of mutual anisotropy of linear and circular birefringence in the diagnostics of oncological changes in morphological structure of biological tissues. Quantum. Electron. **42**, 727–732 (2012)
51. Deboo, B., Sasian, J., Chipman, R.A.: Degree of polarization surfaces and maps for analysis of depolarization. Opt. Exp **12**, 4941–4958 (2004)
52. Cassidy, L.: Basic concepts of statistical analysis for surgical research. J. Surg. Res. **128**(2), 199–206 (2005)
53. Davis, C.S.: Statistical Methods of the Analysis of Repeated Measurements. Springer, New York (2002)
54. Petrie, A., Sabin, C.: Medical Statistics at a Glance. Wiley, Chichester, UK (2009)
55. Buscemi, I.C., Guyot, S.: Near real-time polarimetric imaging system. J. Biomed. Opt. **18**, 116002 (2013)
56. Manhas, S., Vizet, J., Deby, S., Vanel, J.C., Boito, P., Verdier, M., Pagnoux, D.: Demonstration of full 4×4 Mueller polarimetry through an optical fiber for endoscopic applications. Opt. Exp. **23**, 3047–3054 (2015)
57. Pierangelo, A., Manhas, S., Benali, A., Fallet, C., Totobenazara, J.L., Antonelli, M.R., Validire, P.: Multispectral Mueller polarimetric imaging detecting residual cancer and cancer regression after neoadjuvant treatment for colorectal carcinomas. J. Biomed. Opt. **18**, 046014 (2013)
58. Vladimir, Z., Wang, J.B., Yan, X.H.: Human blood plasma crystal and molecular biocolloid textures—dismetabolism and genetic breaches. Nat. Sci. J. Xiangtan Univ. **23**, 118–127 (2011)

Chapter 2
Materials and Methods

2.1 Experimental Setup of Mueller-Polarimeter

Figure 2.1 illustrates optical scheme of classic polarimetry setup [1–6].

The optical sensing of biological sample 6 was realized by parallel (diameter $\emptyset = 104\,\mu m$) beam of the "blue" diode laser 1 (wavelength $\lambda = 0.405\,\mu m$). The light source include the quarterwave plates (elements 3, 5, 8: Achromatic True Zero-Order Waveplates) and polarizer (element 4). Lens 7 (Nikon CFI Achromat P, working distance—30 mm, NA—0.1, magnification—4×) was mounted at the geometric focal length relative to the experimental sample of biological tissue 6. The obtained microscopic image was sampled by a set of pixels CCD-camera (element 10: the Imaging Source DMK 41AU02.AS; resolution of $p \times q = 1280 \times 960$). As a result, the topological elements of the two-dimensional image in a range of 2–2000 μm were recorded.

Polarization filtration of images of samples of biological layers provided a phase filter using the "quarter-wave plate 8—polarizer 9" system.

As a result, an array of input data for PC 11 was formed and 2D of Stokes vector parameters $Sv_{i=1;2;3;4}(p \times q)$ algorithmically calculated:

- illuminator (elements 3–5) forms the set of linear (0°; 45°; 90°) and right—($\otimes$) circularly polarized probing coherent beams;
- for each coherent probe, the transmission axis of the linear polarizer (element 9) was rotated by discrete angles $\Omega = 0°; 90°; 45°; 135°$;
- for each discrete rotation, the 2D map of the linearly polarized projections of the intensity $In_{0;90;45;135}^{0;45;90;\otimes}(p \times q)$ were measured;
- phase shifting element 8 was placed in front of the linear polarizer (element 9);
- the optical of this quarterwave phase late (element 8) was rotated to the discrete angles $\Theta = 45°$ and $\Theta = -45°$;

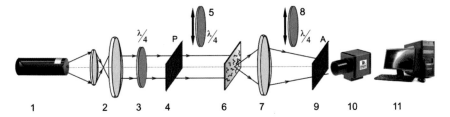

Fig. 2.1 Experimental setup of micropolarimeter: Element 1 is "blue" diode laser; 2 is collimator; 3; 5, 8 quarterwave elements; 4, 9 are linear polarizer's; 6 is biological sample; 7 is polarization lens; 10 is CCD camera; 11 is PC

- the obtained 2D maps of the intensity $In_{\otimes;\oplus}^{0;45;90;\otimes}(q \times p)$ of the polarized-phase-filtered object field were recorded by a digital camera (element 10);

- Stokes-parametric images were calculated $Sv_{i=1;2;3;4}^{0;45;90;\otimes}(q \times p)$:

$$
\begin{pmatrix} Sv_{i=1}^{0;45;90;\otimes} \\ Sv_{i=2}^{0;45;90;\otimes} \\ Sv_{i=3}^{0;45;90;\otimes} \\ Sv_{i=4}^{0;45;90;\otimes} \end{pmatrix} (q \times p) = \begin{pmatrix} In_0^{0;45;90;\otimes} + In_{90}^{0;45;90;\otimes}; \\ In_0^{0;45;90;\otimes} - In_{90}^{0;45;90;\otimes}; \\ In_{45}^{0;45;90;\otimes} - In_{135}^{0;45;90;\otimes}; \\ In_{\otimes}^{0;45;90;\otimes} - In_{\oplus}^{0;45;90;\otimes} \end{pmatrix} (q \times p) \qquad (2.1)
$$

The data obtained (2.1) provided algorithmic obtaining of a set of Mueller-matrix images:

$$
\{W\} = \begin{Vmatrix} W_{11} = 0.5(Sv_1^0 + Sv_1^{90}); & W_{12} = 0.5(Sv_1^0 - Sv_1^{90}); & W_{13} = Sv_1^{45} - W_{11}; & W_{14} = Sv_1^{\otimes} - W_{11}; \\ W_{21} = 0.5(Sv_2^0 + Sv_2^{90}); & W_{22} = 0.5(Sv_2^0 - Sv_2^{90}); & W_{23} = Sv_2^{45} - W_{21}; & W_{24} = Sv_2^{\otimes} - W_{21}; \\ W_{31} = 0.5(Sv_3^0 + Sv_3^{90}); & W_{32} = 0.5(Sv_3^0 - Sv_3^{90}); & W_{33} = Sv_3^{45} - W_{31}; & W_{34} = Sv_3^{\otimes} - W_{31}; \\ W_{41} = 0.5(Sv_4^0 + Sv_4^{90}); & W_{42} = 0.5(Sv_4^0 - Sv_4^{90}); & W_{43} = Sv_4^{45} - W_{41}; & W_{44} = Sv_4^{\otimes} - W_{41}. \end{Vmatrix}
$$

$$(2.2)$$

Using the Eqs. (2.1), (2.2), the elements $\langle\{w_{ik}\}\rangle$ of differential matrix of the 1st order were determined for each pixel of digital camera. Then, using the algorithms, tomograms of the phase $(F(q \times p))$ and amplitude $(D(q \times p))$ anisotropy of fibrillar networks of biological tissues are determined.

Further, we will denote such polarization-phase tomograms as PT.

Algorithmic calculation of maps of the module and phase ($|Sv_{i=3}(q \times p)|$, $Arg(Sv_{i=3}(q \times p))$, $|Sv_{i=4}(q \times p)|$, and $Arg(Sv_{i=4}(q \times p))$) of polarization-inhomogeneous images of biological layers was carried out on the basis of experimental data of Stokes polarimetry [1, 7–13]:

i. The test sample is probed with a circular polarized coherent beam (see Fig. 2.1);

ii. The transmission plane of a linear polarizing filter is discretely oriented by the angles $\Theta = 0°$, $\Theta = 90°$, $\Theta = 45^0$, $\Theta = 135°$. A series of polarized-filtered images $(In_0^\otimes; In_{90}^\otimes; In_{45}^\otimes; In_{135}^\otimes)$ of a sample of a layer of biological tissue is recorded;

iii. In accordance with the algorithm $Sv_{i=1;2;3}^\otimes$—$Sv_1^\otimes = In_0^\otimes + In_{90}^\otimes$; $Sv_2^\otimes = In_0^\otimes - In_{90}^\otimes$; $Sv_3^\otimes = In_{45}^\otimes - In_{135}^\otimes$ a series of linear Stokes-parametric images of the experimental sample are calculated.

iv. The phase-shifting element is placed in front of a linear polarizing filter. Angles $+45°$ and $-45°$ are successively formed between the direction of the optical axis of the phase plate and the transmission axis of the polarizing filter. 2D maps of intensities $In_\otimes^\otimes; In_\oplus^\otimes$ of polarized-filtered images of an optically anisotropic sample of biological tissue are recorded.

v. Based on the algorithm $Sv_4^\otimes = In_\otimes^\otimes - In_\oplus^\otimes$, maps are calculated of the fourth Stokes parameter.

vi. For each partial pixel of a digital camera, the module and phase $(|Sv_{i=3}(q \times p)|, Arg(Sv_{i=3}(q \times p)), |Sv_{i=4}(q \times p)|,$ and $Arg(Sv_{i=4}(q \times p)))$ values of the 3rd and 4th "two-point" parameters of the Stokes vector are calculated:

$$
\begin{cases}
|Sv_3| = \sqrt{\begin{bmatrix} \sqrt{In_0(r_1)In_{90}(r_2)} \cos \delta_2 \\ + \sqrt{In_0(r_2)In_{90}(r_1)} \cos \delta_1 \end{bmatrix}^2 + \begin{bmatrix} \sqrt{In_0(r_1)In_{90}(r_2)} \sin \delta_2 \\ - \sqrt{In_0(r_2)In_{90}(r_1)} \sin \delta_1 \end{bmatrix}^2} ; \\[4em]
ArgSv_3 = arctg \left(\dfrac{\begin{bmatrix} \sqrt{In_0(r_1)In_{90}(r_2)} \sin \delta_2 \\ - \sqrt{In_0(r_2)In_{90}(r_1)} \sin \delta_1 \end{bmatrix}}{\begin{bmatrix} \sqrt{In_0(r_1)In_{90}(r_2)} \cos \delta_2 \\ + \sqrt{In_0(r_2)In_{90}(r_1)} \cos \delta_1 \end{bmatrix}} \right).
\end{cases} \tag{2.3}
$$

$$
\begin{cases}
|Sv_4| = \sqrt{ \begin{aligned} &\left[\begin{array}{l} \sqrt{In_0(r_2)In_{90}(r_1)}\sin\delta_1 \\ + \sqrt{In_0(r_1)In_{90}(r_2)}\sin\delta_2 \end{array} \right]^2 + \\ &+ \left[\begin{array}{l} \sqrt{In_0(r_2)In_{90}(r_1)}\cos\delta_2 + \\ \sqrt{In_0(r_1)In_{90}(r_2)}\cos\delta_1 \end{array} \right]^2 \end{aligned} }; \\[2em]
ArgSv_4 = arctg\left(\dfrac{ \left[\begin{array}{l} \sqrt{In_0(r_2)In_{90}(r_1)}\cos\delta_2 \\ + \sqrt{In_0(r_1)In_{90}(r_2)}\cos\delta_1 \end{array} \right] }{ \left[\begin{array}{l} \sqrt{In_0(r_2)In_{90}(r_1)}\sin\delta_1 \\ + \sqrt{In_0(r_1)In_{90}(r_2)}\sin\delta_2 \end{array} \right] } \right).
\end{cases}
\tag{2.4}
$$

$$
\delta(r) = arctg\left[\left(\frac{Sv_4(r)Sv_2(r)}{Sv_3(r)} \right) \left(\frac{1 + \frac{In_{90}(r)}{In_0(r)}}{1 - \frac{In_{90}(r)}{In_0(r)}} \right) \right].
\tag{2.5}
$$

Here In_0 and In_{90} are the linear components of intensities for the discrete angles of the axis of the linear polarizer $0°$ and $90°$; is the phase shift between the orthogonal linearly polarized components of the amplitude of the laser object field at spatially separated points r_1 and r_2.

2.2 Optical Scheme and 3D Mueller-Matrix Polarimetry Technique

Figure 2.2 shows the 3D polarization-interference scheme of the Stokes-polarimetry of phase-inhomogeneous fields biological layers.

The parallel $\emptyset = 2 \times 10^3\,\mu m$ laser beam (He–Ne laser 1, $\lambda = 0.6328\,\mu m$) has been formed using collimator two. Further, a laser beam has been separated on two

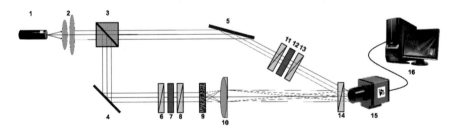

Fig. 2.2 3D interference Stokes polarimeter. Here: 1—coherent probe; 2—optical collimator; 3—amplitude beam splitter; 4; 5—turning mirrors; 6; 8; 11; 13; 14—linear polarizing elements; 7; 12—phase-shifting filters; 9—a sample of biological tissue; 10—polarization objective; 15—CCD; and 16—PC

beams ("irradiating" and "reference") using beam splitter three. The "irradiating" beam has been passed through the polarization filters six–eight using rotary mirror four. Further, this beam illuminated the sample of the biological layer nine.

The "reference" beam has been passed through the polarization filters 11–13 using mirror five. Further, this beam illuminated image plane of the object 9.

The resulting interference pattern passed through the polarizer-analyzer 14 has been recorded using digital camera 15 (The Imaging Source DMK 41AU02.AS).

The polarization states of the "irradiating" and "reference" beams have been formed by means of linear polarization filters 6–8 and 11–13.

Each filter consists of:

- "input" polarizers six and 11 forming plane-polarized beams with azimuth 0°;
- quarterwave plates seven and 12 and polarizer four forming right-circularly polarized beams $\otimes$;
- "output" polarizers eight and 13, forming a series of plane-polarized beams 0°; 90°; 45°.
- The procedure of determination of the layered distributions of diffuse tomo-grams $DT_{red}(\varphi_j; x; y)$ includes the following steps:
- Formation in the irradiating and reference laser beams the next polarization states—0°; 90°; 45°; $\otimes$.
- Filtering the intensity distributions of each interference pattern using a linear polarizing filter 14 for discrete orientations of the transmission axis—$\Omega = 0°$; $\Omega = 90°$.
- Application of Fourier transform procedure to interference pattern and using the inverse Fourier transform to obtaining the distributions of complex amplitudes $U(x,y) = |U|(x,y)\exp i\varphi_j$ in different phase planes ($\varphi_j = \frac{2\pi}{\lambda} r$; $0 \leq r \leq z$) of the object field with an arbitrary step $\Delta\varphi_{j=0...q}$ is realized.
- In each phase plane $\varphi_j(x; y)$, for a series of planar (with azimuths 0°; 90°; 45°) and the right of circularly ($\otimes$) polarized irradiating beams, the distributions of the four sets of parameters of the Stokes vector $Sv_i(0° : 90°; 45°; \otimes)$ are calculated.

$$
\begin{pmatrix}
Sv_1^{(0°;90°;45°;\otimes)}(\varphi_J, x, y) = \left(\left| U_x^{(0°;90°;45°;\otimes)} \right|^2 + \left| U_y^{(0°;90°;45°;\otimes)} \right|^2 \right); \\
Sv_2^{(0°;90°;45°;\otimes)}(\varphi_J, x, y) = \left(\left| U_x^{(0°;90°;45°;\otimes)} \right|^2 - \left| U_y^{(0°;90°;45°;\otimes)} \right|^2 \right); \\
Sv_3^{(0°;90°;45°;\otimes)}(\varphi_J, x, y) = 2\mathrm{Re} \left| U_x^{(0°;90°;45°;\otimes)} U_y^{*(0°;90°;45°;\otimes)} \right|; \\
Sv_4^{(0°;90°;45°;\otimes)}(\varphi_J, x, y) = 2\mathrm{Im} \left| U_x^{(0°;90°;45°;\otimes)} U_y^{*(0°;90°;45°;\otimes)} \right|.
\end{pmatrix} \quad (2.6)
$$

- On the basis of Eq. (2.6) layered Muller-matrix images $W_{ik}(\varphi_j, x, y)$ are determined as:

$$\{W\}(\varphi_j, x, y) = \begin{Vmatrix} W_{11}; & W_{12}; & W_{13}; & W_{14}; \\ W_{21}; & W_{22}; & W_{23}; & W_{24}; \\ W_{31}; & W_{23}; & W_{24}; & W_{34}; \\ W_{41}; & W_{24}; & W_{34}; & W_{44} \end{Vmatrix}(\varphi_j, x, y)$$

$$= 0.5 \begin{pmatrix} \begin{Vmatrix} (Sv_1^0 + Sv_1^{90}); & (Sv_1^0 - Sv_1^{90}); & (Sv_1^{45} - Sv_1^{135}); & (Sv_1^\otimes - Sv_1^\oplus); \\ (Sv_2^0 + Sv_2^{90}); & (Sv_2^0 - Sv_2^{90}); & (Sv_2^{45} - Sv_2^{135}); & (Sv_2^\otimes - Sv_2^\oplus); \\ (Sv_3^0 + Sv_3^{90}); & (Sv_3^0 - Sv_3^{90}); & (Sv_3^{45} - Sv_3^{135}); & (Sv_3^\otimes - Sv_3^\oplus); \\ (Sv_4^0 + Sv_4^{90}); & (Sv_4^0 - S_4^{90}); & (Sv_4^{45} - Sv_4^{135}); & (Sv_4^\otimes - Sv_4^\oplus) \end{Vmatrix} \end{pmatrix}(\varphi_j, x, y)$$

$$(2.7)$$

Based on the set of distributions (2.7), a series of layer wise distributions of phase anisotropy $W_{ik}(\varphi_j; x; y) \rightarrow 0.5\,r^{-2}(\ln W_{22} + \ln W_{33} + \ln W_{44})^{-1}$

$$\left\{\begin{array}{l} 1 \\ \ln W_{22}\left(B\,\tilde{L}_{0;90}\right); \\ \ln W_{33}\left(B\,\tilde{L}_{45;135}\right); \\ \ln W_{44}\left(B\,\tilde{C}_{\otimes;\oplus}\right) \end{array}\right\}(\varphi_j; x; y) \text{ is determined.}$$

Statistical Analysis

The algorithm for the statistical processing of arrays of the magnitude of the module and phase $q = \left\{\begin{array}{l} |Sv_{i=3;4}(q \times p)| \\ Arg\left(Sv_{i=3;4}(q \times p)\right) \end{array}\right.$ of the polarization-correlation maps of "two-point" parameters of the Stokes vector is illustrated by the relations [3]:

$$Z_1 = \frac{1}{N}\sum_{i=1}^{N}|(q)_i|; Z_2 = \sqrt{\frac{1}{N}\sum_{i=1}^{N}(q)_i^2};$$

$$Z_3 = \frac{1}{Z_2^3}\frac{1}{N}\sum_{i=1}^{N}(q)_i^3; Z_4 = \frac{1}{Z_2^4}\frac{1}{N}\sum_{i=1}^{N}(q)_i^4. \tag{2.8}$$

Here, N is the number (1280×960) of pixel at the CCD.

Correlation Analysis

To characterize the coordinate uniformity of the polarization-correlation maps of the "two-point" parameters of the Stokes vector, the autocorrelation functions $\bar{K}(\Delta x)$ with scanning step $\Delta x = 1pix$ were calculated in rows of CCD-camera pixels [2]:

$$AK_{j=1 \div n}(\Delta x) = \lim_{x \to 0}\frac{1}{m}\int_0^x \left[q_j^*(x = 1 \div m)\right][q^*(x - \Delta x)]dx. \tag{2.9}$$

Here, $\bar{q}$ is the average value of the set q.

$$A\bar{K}(\Delta x) = \sum_{j=1}^{n} \frac{AK_j(\Delta x)}{n}. \tag{2.10}$$

Fractal Analysis

The degree of scale self-similarity of the polarization-correlation maps of the "two-point" parameters of the Stokes vector was calculated using the logarithmic dependences $(\log S(q) - \log(v))$ of power spectra of random values of the module and distributions phase. The obtained distributions $\log S(q) - \log(v)$ are algorithmically (the least squares method) transformed into approximating curves $\Phi(\eta)$. The obtained types of such curves determine [1]:

- linear with one slope $\Phi(\eta)$ (for 2–3 decades of changing the size of structural elements), the distribution of the modulus and phase of the "two-point" parameters of the Stokes vector are fractal;
- linear with several slopes of the $\Phi(\eta)$—distribution of the module and phase of the "two-point" parameters of the Stokes vector are multifractal;
- nonlinear $\Phi(\eta)$ (for 2–3 decades of changing the size of structural elements) - distributions of the module and phase of the "two-point" parameters of the Stokes vector are statistical;

For a quantitative characteristic of the logarithmic distributions $\log S(q) - \log(v)$ was introduced a statistical moment of the 2nd order:

$$D^f = \sqrt{\frac{1}{N} \sum_{j=1}^{N} \left(\log S(q) - \log(v)^2 \right)_j}. \tag{2.11}$$

References

1. Ushenko, A., Pishak, V.: Laser polarimetry of biological tissue: principles and applications. In: Tuchin, V. (ed.) Handbook of Coherent-Domain Optical Methods: Biomedical Diagnostics, Environmental and Material Science, pp. 93–138 (2004)
2. Angelsky, O., Ushenko, A., Ushenko, Y., Pishak, V., Peresunko, A.: Statistical, correlation and topological approaches in diagnostics of the structure and physiological state of birefringent biological tissues. In: Handbook of Photonics for Biomedical Science, pp. 283–322 (2010)
3. Ushenko, Y., Boychuk, T., Bachynsky, V., Mincer, O.: Diagnostics of structure and physiological state of birefringent biological tissues: statistical, correlation and topological approaches. In: Tuchin, V (ed.) Handbook of Coherent-Domain Optical Methods. Springer (2013)
4. Ushenko, V.A., Dubolazov, A.V.: Correlation and self similarity structure of polycrystalline network biological layers Mueller matrices images. Proc. SPIE **8856**, 88562D (2013)

5. Ushenko, V.O.: Spatial-frequency polarization phasometry of biological polycrystalline networks. Opt. Mem. Neur. Netw. **22**, 56–64 (2013)
6. Ushenko, V.A., Pavlyukovich, N.D., Trifonyuk, L.: Spatial-frequency azimuthally stable cartography of biological polycrystalline networks. Int. J. Opt. **683174**, 2013 (2013)
7. Chipman, R.: Polarimetry. In: Bass, M. (eds.) Handbook of Optics: Vol. I—Geometrical and Physical Optics, Polarized Light, Components and Instruments, pp. 22.1–22.37. McGraw-Hill Professional, New York (2010)
8. Ghosh, N., Wood, M., Vitkin, A.: Polarized light assessment of complex turbid media such as biological tissues via Mueller matrix decomposition. In: Tuchin V (ed.) Handbook of Photonics for Biomedical Science, pp. 253–282. CRC Press, Taylor & Francis Group, London (2010)
9. Jacques, S.: Polarized light imaging of biological tissues. In: Boas, D., Pitris, C., Ramanujam, N. (eds.) Handbook of Biomedical Optics, pp. 649–669. CRC Press, Boca Raton, London, New York (2011)
10. Ghosh, N.: Tissue polarimetry: concepts, challenges, applications, and outlook. J. Biomed. Opt. **16**(11), 110801 (2011)
11. Swami, M., Patel, H., Gupta, P.: Conversion of 3×3 Mueller matrix to 4×4 Mueller matrix for non-depolarizing samples. Opt. Commun. **286**, 18–22 (2013)
12. Lu, S.Y., Chipman, R.A.: Interpretation of Mueller matrices based on polar decomposition. J. Opt. Soc. Am. A **13**, 1106–1113 (1996)
13. Guo, Y., Zeng, N., He, H., Yun, T., Du, E., Liao, R., Ma, H.: A study on forward scattering Mueller matrix decomposition in anisotropic medium. Opt. Exp. **21**, 18361–18370 (2013)

Chapter 3
Results and Discussion

This part of the research направлена on the discussion of the results of the used methods.

- Polarization-phase tomography of the distributions of linear and circular birefringence and dichroism of polycrystalline networks of biological tissues samples.
- Mueller-matrix-based polarization imaging and quantitative assessment of optically anisotropic polycrystalline networks.
- Polarization reconstruction of the distribution of the magnitude of the fluctuations of linear and circular birefringence and dichroism of myocardium layers.
- 3D layered distribution maps of the fluctuations of linear and circular birefringence and dichroism of polycrystalline blood films for cancer diagnosis.
- Stokes-correlometry analysis of biological tissues with polycrystalline structure

3.1 Polarization-Phase Matrix Reconstruction of the Structural Anisotropy Distributions of Fibrillar Networks of Biological Tissues

The results of a study of the diagnostic effectiveness of polarization-phase tomography of the distributions of linear and circular birefringence and dichroism in the forensic medical assessment of degenerative-dystrophic changes in myocardial tissue are presented.

V. T. Bachinskyi et al., *Polarization Correlometry of Scattering Biological Tissues and Fluids*, SpringerBriefs in Physics, https://doi.org/10.1007/978-981-15-2628-2_3

3.1.1   Samples Preparation

Biological preparations of both types of tissues were made by the traditional method. A biopsy of myocardium was placed on the microtome work table and underwent rapid freezing. Thin sections (geometrical thickness $d = 30\,\mu m \div 40\,\mu m$) were cut from the obtained myocardial samples. Then they were placed on optically uniform glass. Thus, native histological sections were formed with the following optical parameters:

- attenuation coefficient $\tau = 0.045 \div 0.075$;
- degree of depolarization $\Lambda = 34\% \div 43\%$).

The following representative samples of samples were formed: native sections of myocardial biopsy of the patients who died of acute coronary insufficiency (control type A, 36 samples) and ischemic disease (research type B, 36 samples).

3.1.2   Matrix Reconstruction of the Phase and Amplitude Anisotropy of Myocardium Layers

The combination of linear and circular birefringence and dichroism tomograms, as

well as histograms of the distributions of the magnitude $T(\langle k_{ik} \rangle) = \begin{Bmatrix} \delta \\ \varphi \\ \tau \\ \chi \end{Bmatrix}(x, y)$ for

the samples of the myocardium are presented in the series of Figs. 3.1 and 3.2.

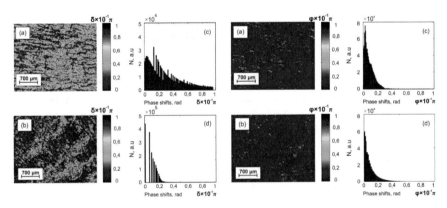

Fig. 3.1 Maps ((1), (3), (5), (7)) and histograms ((2), (4), (6), (8)) of the distributions of linear $\delta(x, y)$ and circular $\varphi(x, y)$ birefringence of myocardium samples from type A ((1)–(4)) and type B ((5)–(8))

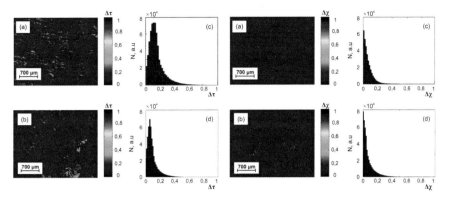

Fig. 3.2 Maps ((1), (3), (5), (7)) and histograms ((2), (4), (6), (8)) of the distributions of linear $\tau(x, y)$ and circular $\chi(x, y)$ birefringence of myocardium samples from type A ((1)–(4)) and type B ((5)–(8))

Analysis of the data showed:

- The presence of phase and amplitude anisotropy of myocardial samples from groups A and B.

- Distributions of magnitude $T(\langle k_{ik} \rangle) = \begin{Bmatrix} \delta \\ \varphi \\ \tau \\ \chi \end{Bmatrix}(x, y)$ are characterized by a significant average and interval of variation of linear and circular birefringence and dichroism of fibrillar networks.

  - The prevalence of polarization manifestations of the phase structural (the linear birefringence) anisotropy (fragments ((1), (2), (5), (6), Fig. 3.1)) over optical activity (fragments ((3), (4), (7), (8), Fig. 3.1)) and optically anisotropic absorption (fragments ((1)–(8), Fig. 3.2)) anisotropy.

- A higher level of linear birefringence of the type A myocardial fibrillar networks in comparison with samples of type B.

From a physical point of view, the revealed differences in the polarization manifestations of the parameters of phase and amplitude anisotropy can be associated with the following factors. Firstly—the spectral range ("red", $\lambda = 0.64\,\mu m$) of Mueller-matrix tomography. In this long-wavelength region of the spectrum of visible radiation, absorption by fibrillar protein networks is minimal. Therefore, for myocardial samples of all types, the predominance ($\begin{Bmatrix} \langle \delta \rangle > \langle \tau \rangle; \\ \langle \varphi \rangle > \langle \chi \rangle \end{Bmatrix}$) of phase takes place above the amplitude anisotropy, the maximum level of which is realized in the ultraviolet and blue spectral regions. Secondly, the destruction of structural anisotropy (myocardium from group B) due to degenerative-dystrophic changes in myosin networks (disordering and thinning of partial fibrils) leads to a decrease in the level of linear birefringence and dichroism. Quantitatively, this scenario illustrates the transformation (decrease in the average and the range of variation) of histograms of distributions (fragments ((2), (6), Fig. 3.1)) and (fragments ((2), (6), Fig. 3.2)).

3.2 Mueller-Matrix-Based Polarization Imaging and Quantitative Assessment of Optically Anisotropic Polycrystalline Networks

We introduce a new Mueller-matrix-based polarization approach for quantitatively measureable mapping of polycrystalline structure of histological sections of fibrillar biological tissues. The high-order statistical moments of spatial distributions of linear and circular birefringence, dichroism and their variations are utilized for quantitative non-invasive assessment of the layers of myocardium, endometrium and connective tissue component of the ligament of the uterus. We show that distribution of phase and amplitude anisotropy of biological tissues at дегенеративно-дистрофических and pathological stages can be used as the diagnostic parameters. Thus, based on the statistical analysis of карт оптической анизотропии of fibrillar networks of biological tissues the differentiation criteria of necrotic conditions of the myocardium, malignant changes in the endometrium and connective tissue component of the ligament of the uterus during genital prolapse were defined.

3.2.1 Samples, Preparation and Statistical Validation

We investigated samples of fibrillar tissue of human organs with difficult to diagnose necrotic and pathological conditions:

- myocardium of deceased patients as a result of ischemic heart disease (IHD—group 1 "control") and acute coronary insufficiency (ACI—group 2 "diagnosed");
- early oncological conditions ("Norm"—group 3) and cancer of the first stage ("Cancer"—group 4) of the endometrium;
- pathological changes in the connective tissue component of the ligament of the uterus with prolapse of genitalia (group 5—healthy areas ("Normal"), group 6—pathologically changed area ("Pathology") operatively extracted wall of the vagina).

The objects selected for the study combine the similarity of a polycrystalline structure—the presence in all cases of fibrillar networks $(\Delta n_{0;90}; \Delta n_{45;135}; \Delta \mu_{0;90}; \mu_{45;135})$ formed by optically active $(\Delta n_{\otimes;\oplus}; \Delta \mu_{\otimes;\oplus})$ protein molecules of myosin and collagen. A comparative qualitative analysis of microscopic images revealed (see Fig. 3.3):

- individual structure of polarization-visualized fibrillar networks of various biological tissues;
- absence of pronounced differences between the polycrystalline structures of all tissues within the control and investigated groups.

The differentiation of necrotic and pathological conditions of biological tissues of all types was performed by the biopsy of surgically removed samples, which is believed to be a gold standard method.

3.2.2   Stokes Parametric Mapping of Biological Tissues Layers

Figures 3.4, 3.5 and 3.6 present the polarization tomograms $PT(p \times q)$ and the histograms $N(PT)$ of the distribution of linear birefringence $F(p \times q)$ and dichroism $D(p \times q)$ anisotropy of the myocardium (Fig. 3.4), endometrium (Fig. 3.4) and connective tissue component of the ligament of the uterus (Fig. 3.6) histological sections.

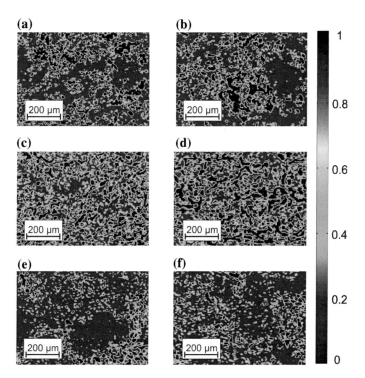

Fig. 3.3 Microscopic (4X) polarization-visualized images of the поликристаллической компонent of the нативных слоев of the myocardium (**a, b**), endometrium (**c, d**) and connective tissue component of the uterine ligament (**e, f**) of control (**a, c, e**) and investigated (**b, d, f**) groups

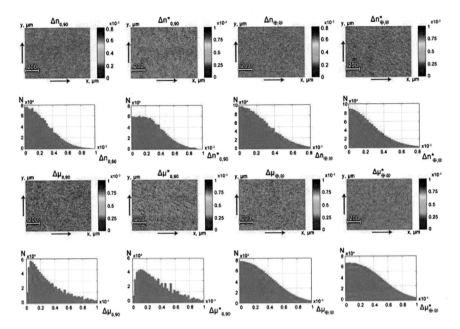

Fig. 3.4 Polarization-phase tomograms **PT** and histograms N of the distributions $F(p \times q)$ and $D(p \times q)$ of myocardium samples of deceased patients from group 1 ($\Delta n_{0;90}$, $\Delta n_{\otimes;\oplus}$, $\Delta \mu_{0;90}$, $\Delta \mu_{\otimes;\oplus}$) and group 2 ($\Delta n^*_{0;90}$, $\Delta n^*_{\otimes;\oplus}$, $\Delta \mu^*_{0;90}$, $\Delta \mu^*_{\otimes;\oplus}$)

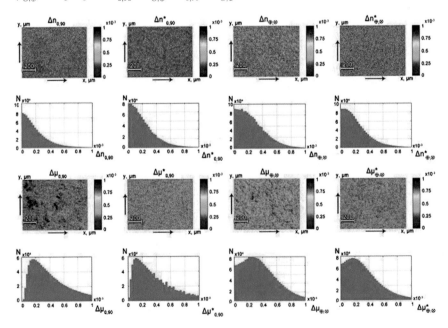

Fig. 3.5 Polarization-phase tomograms **PT** and histograms N of the distributions $F(p \times q)$ and $D(p \times q)$ of endometrium histological sections of patients from group 3 ($\Delta n_{0;90}$, $\Delta n_{\otimes;\oplus}$, $\Delta \mu_{0;90}$, $\Delta \mu_{\otimes;\oplus}$) and group 4 ($\Delta n^*_{0;90}$, $\Delta n^*_{\otimes;\oplus}$, $\Delta \mu^*_{0;90}$, $\Delta \mu^*_{\otimes;\oplus}$)

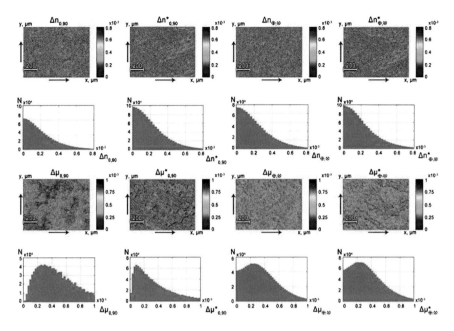

Fig. 3.6 Polarization-phase tomograms **PT** and histograms N of the distributions $F(p \times q)$ and $D(p \times q)$ of connective tissue component histological sections of patients from group 5 ($\Delta n_{0;90}$, $\Delta n_{\otimes;\oplus}$, $\Delta \mu_{0;90}$, $\Delta \mu_{\otimes;\oplus}$) and group 6 ($\Delta n^*_{0;90}$, $\Delta n^*_{\otimes;\oplus}$, $\Delta \mu^*_{0;90}$, $\Delta \mu^*_{\otimes;\oplus}$)

Myocardium

The linear birefringence $\Delta n_{0;90}$ and dichroism $\Delta \mu_{0;90}$ have a higher level in comparison with the circular phase $\Delta n_{\otimes;\oplus}$ and amplitude $\Delta \mu_{\otimes;\oplus}$ anisotropy (see Fig. 3.4). Quantitatively this manifests itself in lower values of main extrema of $N(\Delta n_{0;90}) = 0$ and $N(\Delta \mu_{0;90} = 0)$. Due to this, larger average values

$$\left(\begin{cases} \bar{\Delta} n_{0;90} > \bar{\Delta} n_{\otimes;\oplus}; \\ \bar{\Delta} \mu_{0;90} > \bar{\Delta} \mu_{\otimes;\oplus} \end{cases} \right)$$ of histograms $N(\Delta n_{0;90})$ and $N(\Delta \mu_{0;90})$ are formed. The

comparison of the myocardium samples polarization phase tomograms has lower level of linear birefringence and dichroism in control group 1 (IHD)

$$- \begin{cases} \bar{\Delta} n_{0;90}(\textbf{IHD}) < \bar{\Delta} n^*_{0;90}(\textbf{ACI}); \\ \bar{\Delta} \mu_{0;90}(\textbf{IHD}) < \bar{\Delta} \mu^*_{0;90}(\textbf{ACI}) \end{cases} \text{ (see Fig. 3.4).}$$

The obtained results are believed to be conditioned by the following physical considerations. Long process of IHD leads to degenerative and dystrophic changes of myocardium. As a result, the level of structural anisotropy decreases and the histograms $N(\Delta n_{0;90})$, $N\left(\Delta n^*_{0;90}\right)$ and $N(\Delta \mu_{0;90})$, $N\left(\Delta \mu^*_{0;90}\right)$ (see Fig. 3.4) are characterized by lower values of main extrema of $N(\Delta n_{0;90}) = 0$ and $N(\Delta \mu_{0;90} = 0)$. In addition, the average (Z_1), half-width (dispersion Z_2), as well as skewness (Z_3) and sharpness of the peak (kurtosis Z_4) of such dependences are different.

Endometrium

As in the case of the myocardium, the magnitude of the linear birefringence and dichroism of the fibrillar collagen network is greater than the circular phase and amplitude anisotropy (see Fig. 3.4). Differences between average values and ranges of distributions $\begin{cases} \Delta n_{0;90}(p \times q); \\ \Delta n_{\otimes;\oplus}(p \times q) \end{cases}$ and $\begin{cases} \Delta \mu_{0;90}(p \times q); \\ \Delta \mu_{\otimes;\oplus}(p \times q) \end{cases}$, determined for endometrial samples, are less pronounced than for myocardium tissue (Fig. 3.4).

Differences between average values and ranges of distributions $\begin{cases} \Delta n_{0;90}(p \times q); \\ \Delta n_{\otimes;\oplus}(p \times q) \end{cases}$

and $\begin{cases} \Delta \mu_{0;90}(p \times q); \\ \Delta \mu_{\otimes;\oplus}(p \times q) \end{cases}$, determined for endometrial samples, are less pronounced than for myocardium tissue. This fact can be related to the smaller spatial structure and geometric dimensions of collagen fibrils of the endometrium in comparison with the myocardium fibrillar network.

A comparative analysis of the polarization-phase tomograms of endometrial samples revealed a decrease in the level of linear birefringence and dichroism of the fibrillar network of collagen fibers of the tissue with malignant changes $-\begin{cases} \bar{\Delta} n *_{0;90} (\mathbf{Cancer}) < \bar{\Delta} n_{0;90}(\mathbf{Norm}); \\ \bar{\Delta} \mu *_{0;90} (\mathbf{Cancer}) < \bar{\Delta} \mu_{0;90}(\mathbf{Norm}) \end{cases}$ (see Fig. 3.5). The onset of the pathological process leads to necrotic changes in the structure (disorientation and reduction in size) of the fibrillar network of the endometrium. Therefore, as in the case of myocardium, the distribution histograms of the value of linear birefringence $N(\Delta n*_{0;90})$ and the dichroism $H(\Delta \mu*_{0;90})$ (see Fig. 3.5) obtained for histological sections of endometrium from group 4 are characterized by lower values of average (Z_1) and half-width (dispersion Z_2). On the contrary, the skewness (Z_3) and the sharpness of the peak (kurtosis Z_4) of such dependences increase.

Connective Tissue Component of the Ligament of the Uterus

For this tissue, the magnitude of the linear birefringence and dichroism is commensurate with the mechanisms of the circular phase and amplitude anisotropy $\begin{cases} \bar{\Delta} n_{0;90}(\mathbf{normal}) \approx \bar{\Delta} n_{\otimes;\oplus}(\mathbf{normal}); \\ \bar{\Delta} \mu_{0;90}(\mathbf{normal}) \approx \bar{\Delta} \mu_{\otimes;\oplus}(\mathbf{normal}) \end{cases}$ and $\begin{cases} \bar{\Delta} n *_{0;90} (\mathbf{pathology}) \approx \bar{\Delta} n *_{\otimes;\oplus} (\mathbf{pathology}); \\ \bar{\Delta} \mu *_{0;90} (\mathbf{pathology}) \approx \bar{\Delta} \mu *_{\otimes;\oplus} (\mathbf{pathology}) \end{cases}$ (see Fig. 3.6).

The observed fact can be associated with another morphological structure of connective tissue—the absence of spatial ordering of collagen fibers in comparison with more structured myocardium fibrillar networks and endometrium.

Comparative analysis of the optically anisotropic structure of samples of the connective tissue component of uterine ligaments from group 5 ("normal") and group 6 ("pathology"—degenerative-dystrophic changes) reveals a decrease in the level of phase ($\Delta n_{0;90}; \Delta n_{\otimes;\oplus}$) and amplitude ($\Delta \mu_{0;90}; \Delta \mu_{\otimes;\oplus}$) anisotropy of pathologically changed tissue. Polarized it is manifested in the decrease of linear

$$\begin{cases} \bar{\Delta} n *_{0;90} (\textbf{Pathology}) < \bar{\Delta} n_{0;90}(\textbf{Normal}); \\ \bar{\Delta} \mu *_{0;90} (\textbf{Pathology}) < \bar{\Delta} \mu_{0;90}(\textbf{Normal}) \end{cases}$$ and circular

$$\begin{cases} \bar{\Delta} n *_{\otimes;\oplus} (\textbf{Pathology}) \prec \bar{\Delta} n_{\otimes;\oplus}(\textbf{Normal}); \\ \bar{\Delta} \mu *_{\otimes;\oplus} (\textbf{Pathology}) \prec \bar{\Delta} \mu_{\otimes;\oplus}(\textbf{Normal}) \end{cases}$$ birefringence   and   dichroism   (see

Fig. 3.6).

Statistical Analysis

In this part, the diagnostic effectiveness of data from two methods—polarization microscopy ($In_{0;90}(m \times n)$) and polarization-phase tomography ($F(p \times q)$ and $D(p \times q)$) is analyzed within the statistical approach.

We used the traditional method of statistical differentiation of two representative samples—groups of control ("1"; "3"; "5") and studied groups ("2"; "4"; "6") [1, 2]:

- within each group, the average values $\tilde{Z}_{i=1;2;3;4}$ of the values of each of the statistical moments of the 1st–4th orders $Z_{i=1;2;3;4}$ and standard deviation $\sigma_{i=1;2;3;4}$ were determined;
- if the value $\tilde{Z}_{i=1;2;3;4}$ taking into account standard deviation $\sigma_{i=1;2;3;4}$ within of groups of control ("1"; "3"; "5") does not match with the same value within the investigated groups ("2"; "4"; "6") and vice versa, then the differences between the statistical parameters are reliable;
- within the representative samples (3σ) of both groups biological samples we determined the number of "false negative" (b) and "false positive" (d) results;
- the strength characteristics of the method were calculated [3]—sensitivity ($Se = \frac{a}{a+b} 100\%$), specificity ($Sp = \frac{c}{c+d} 100\%$) and balanced accuracy ($Ac = \frac{Se+Sp}{2}$), where a and b are the number of correct and wrong diagnoses within group ("1"; "3"; "5"); c and d—the same within group ("2"; "4"; "6").

Method of Polarization Microscopy

The results of the statistical and information analysis of the intensity distributions of the polarization-visualized images of the polycrystalline structure of biological tissues of all types are presented in Table 3.1.

The obtained results show insufficient level of accuracy of polarization microscopy of degenerative-dystrophic conditions in tissues of all types. The value of balanced accuracy does not exceed 70%.

Table 3.1 Results of the statistical and information analysis of polarization images of the оптической анизотропии of biological tissues

	Miocardium			Endometrium			Connective tissue		
	Group 1	Group 2	Ac (%)	Group 1	Group 2	Ac (%)	Group 1	Group 2	Ac (%)
Z_1	0.33 ± 0.019	0.25 ± 0.017	68.5	0.23 ± 0.015	0.19 ± 0.011	62.5	0.15 ± 0.009	0.11 ± 0.006	66.5
Z_2	0.24 ± 0.016	0.18 ± 0.011	70.5	0.18 ± 0.011	0.16 ± 0.009	60.5	0.14 ± 0.008	0.09 ± 0.005	67.8
Z_3	0.36 ± 0.022	0.43 ± 0.025	66.5	0.42 ± 0.023	0.55 ± 0.033	64.5	0.54 ± 0.032	0.68 ± 0.037	65.5
Z_4	0.44 ± 0.025	0.57 ± 0.034	67.5	0.63 ± 0.035	0.74 ± 0.042	63.5	0.76 ± 0.043	0.88 ± 0.052	64.5

Table 3.2 Average $\tilde{Z}_{i=1;2;3;4}$ and standard deviations of the magnitude of statistical moments of the 1st–4th orders that characterize the distribution of the anisotropy parameters of the fibrillar networks of myocardium

Parameters	$\Delta n_{0;90}$	$\Delta^* n_{0;90}$	$\Delta n_{\otimes;\oplus}$	$\Delta^* n_{\otimes;\oplus}$	$\Delta \mu_{0;90}$	$\Delta \mu^*_{0;90}$	$\Delta \mu_{\otimes;\oplus}$	$\Delta \mu^*_{\otimes;\oplus}$
	Group 1	Group 2	Group 1	Group 2	Group 1	Group 2	Group 1	Group 2
$Z_1 \times 10^{-3}$	0.22 ± 0.02	0.29 ± 0.016	0.00026 ± 0.00016	0.004 ± 0.00016	0.008 ± 0.0004	0.006 ± 0.0003	0.0047 ± 0.00025	0.0041 ± 0.00023
$Z_2 \times 10^{-3}$	0.54 ± 0.026	0.43 ± 0.021	0.45 ± 0.023	0.37 ± 0.024	0.91 ± 0.062	0.72 ± 0.041	0.75 ± 0.043	0.63 ± 0.032
Z_3	**0.38 ± 0.022**	**0.57 ± 0.033**	0.49 ± 0.031	0.43 ± 0.024	**0.72 ± 0.037**	**1.09 ± 0.69**	0.29 ± 0.016	0.36 ± 0.019
Z_4	**0.52 ± 0.027**	**0.82 ± 0.044**	0.68 ± 0.036	0.55 ± 0.029	**0.45 ± 0.025**	**0.74 ± 0.038**	0.18 ± 0.011	0.24 ± 0.015

Method of Polarization-Phase Tomography

Myocardium

Table 3.2 shows the results of a statistical analysis of the data of polarization reconstruction of the distributions of linear and circular birefringence and the dichroism of the optically anisotropic component of representative samples of biological tissues from group 1 and group 2. The comparative analysis of the data obtained showed that the differences between the value.

A comparative analysis of the results shown in Table 3.2 revealed:

- varying degrees of sensitivity of statistical moments of the 1st–4th orders to the intergroup differentiation of phase anisotropy distributions;
- the values of the overlap of the distribution values $N(Z_i)$ are individual for various statistical moments and parameters of various mechanisms of optical anisotropy;
- the highest level of overlap occurs for the average distributions of linear birefringence and dichroism, and statistical moments of the 1st to 4th order, which characterize the maps of circular birefringence and dichroism;
- the statistical moments $Z_{i=2;3;4}(\Delta n_{0;90})$ and $Z_{i=3;4}(\Delta \mu_{0;90})$ were the most sensitive to changes in the optical anisotropy of myocardial fibrillar networks with a minimum level of overlap of eigenvalues.

The results of a statistical analysis of a set of polarization-phase tomograms within representative samples of myocardial samples with various degenerative-dystrophic changes were used to determine the operational characteristics of this method.

Table 3.3 illustrates the balanced accuracy of the polarization reconstruction method for the distributions of the phase and amplitude anisotropy parameters of the myocardial layers from group 1 and group 2.

High levels of balanced accuracy of matrix tomography techniques have been established:

Table 3.3 Accuracy of the method of polarization-phase reconstruction of optical anisotropy maps of myocardial fibrillar networks

Parameters	Z_i	$\Delta n_{0;90}$ (%)	$\Delta n_{\otimes;\oplus}$ (%)	$\Delta \mu_{0;90}$ (%)	$\Delta \mu_{\otimes;\oplus}$ (%)
$Ac(Z_i)$	Z_1	86.4	62.3	86.3	61.3
	Z_2	88.2	65.4	68.4	65.4
	Z_3	**91.3**	78.2	**89.2**	72.3
	Z_4	**95.2**	82.3	**94.3**	69.2

Table 3.4 Average $\tilde{Z}_{i=1;2;3;4}$ and standard deviations of the statistical moments of the 1st—4th orders of magnitude, which characterize the distribution of the anisotropy parameters of the fibrillar networks of endometrium

Parameters	$\Delta n_{0;90}$	$\Delta^* n_{0;90}$	$\Delta n_{\otimes;\oplus}$	$\Delta^* n_{\otimes;\oplus}$	$\Delta \mu_{0;90}$	$\Delta \mu^*_{0;90}$	$\Delta \mu_{\otimes;\oplus}$	$\Delta \mu^*_{\otimes;\oplus}$
	Group 3	Group 4	Group 3	Group 4	Group 3	Group 4	Group 3	Group 4
$Z_1 \times 10^{-3}$	0.31 ± 0.024	0.22 ± 0.015	0.16 ± 0.071	0.21 ± 0.011	0.71 ± 0.051	0.46 ± 0.024	0.36 ± 0.023	0.42 ± 0.025
$Z_2 \times 10^{-3}$	0.41 ± 0.022	0.32 ± 0.013	0.33 ± 0.014	0.36 ± 0.021	0.72 ± 0.036	0.52 ± 0.041	0.43 ± 0.024	0.51 ± 0.032
Z_3	**0.48 ± 0.026**	**0.69 ± 0.037**	0.56 ± 0.033	0.48 ± 0.027	**0.98 ± 0.046**	**1.37 ± 0.072**	0.38 ± 0.022	0.45 ± 0.028
Z_4	**0.62 ± 0.036**	**0.92 ± 0.053**	0.75 ± 0.044	0.62 ± 0.036	**0.55 ± 0.031**	**0.88 ± 0.049**	0.2 ± 0.019	0.34 ± 0.022

- asymmetry of the distributions of linear birefringence and dichroism $Ac(Z_3(\Delta n_{0;90}; \Delta \mu_{0;90})) \sim 90\%$—good level;
- excess of the distributions of linear birefringence and dichroism $Ac(Z_4(\Delta n_{0;90}; \Delta \mu_{0;90})) \succ 90\%$—high quality.

Endometrium

The results of similar studies of statistical reliability and diagnostic effectiveness of the Mueller-matrix reconstruction technique of the polycrystalline structure of endometrial biological layers are presented in Tables 3.4 and 3.5.

The analysis of the obtained data of the polarization-phase reconstruction of the polycrystalline structure of endometrial samples showed:

- statistical moments of higher orders (asymmetry Z_3 and excess Z_4)—are most sensitive to pathological changes in the distributions of linear birefringence $(\Delta n_{0;90})$ and dichroism $(\Delta \mu_{0;90})$ (depicted by bold—Table 3.4);
- high level of balanced accuracy of the method in the diagnosis of preclinical stages of endometrial cancer—good $(Ac(Z_3(\Delta \mu_{0;90}); Z_4(\Delta n_{0;90})) \sim 90\%)$ and excellent $(Ac(Z_3(\Delta n_{0;90}); Z_4(\Delta \mu_{0;90})) > 90\%)$ accuracy (Table 3.5).

Connective Tissue Component of the Ligament of the Uterus

The series of Tables 3.6 and 3.7 shows the data of statistical reliability and strength of the method of polarization matrix reconstruction of the polycrystallite component of the samples of the biopsy of the connective tissue component of the ligament of the uterus during genital prolapsed.

The results of the polarization-phase reconstruction of the polycrystalline structure of the samples of the connective tissue component of the uterine ligament revealed:

- maximum sensitivity of statistical moments of the 3rd and 4th orders (asymmetry Z_3 and excess Z_4) to changes in the linear and circular phase and amplitude anisotropy of the connective tissue component of the uterine ligament during the prolapse of the genitalia (depicted by bold—Table 3.6);

Table 3.5 The accuracy of the polarization-phase reconstruction method of optical anisotropy maps of endometrium

Parameters	Z_i	$\Delta n_{0;90}$ (%)	$\Delta n_{\otimes;\oplus}$ (%)	$\Delta \mu_{0;90}$ (%)	$\Delta \mu_{\otimes;\oplus}$ (%)
$Ac(Z_i)$	Z_1	82.2	66.3	84.3	64.3
	Z_2	84.3	68.2	86.4	66.2
	Z_3	**92.1**	74.4	**87.2**	78.4
	Z_4	**88.4**	76.2	**92.4**	72.2

Table 3.6 Average $\tilde{Z}_{i=1;2;3;4}$ and standard deviations that characterize the distribution of the anisotropy parameters of connective tissue component of the ligament of the uterus

Parameters	$\Delta n_{0;90}$	$\Delta^* n_{0;90}$	$\Delta n_{\otimes;\oplus}$	$\Delta^* n_{\otimes;\oplus}$	$\Delta \mu_{0;90}$	$\Delta \mu^*_{0;90}$	$\Delta \mu_{\otimes;\oplus}$	$\Delta \mu^*_{\otimes;\oplus}$
	Group 5	Group 6	Group 5	Group 6	Group 5	Group 5	Group 5	Group 6
$Z_1 \times 10^{-3}$	0.24 ± 0.011	0.14 ± 0.008	0.1 ± 0.009	0.22 ± 0.012	0.41 ± 0.022	0.34 ± 0.017	0.31 ± 0.019	0.33 ± 0.018
$Z_2 \times 10^{-3}$	0.32 ± 0.022	0.26 ± 0.013	0.33 ± 0.015	0.36 ± 0.021	0.53 ± 0.033	0.44 ± 0.021	0.36 ± 0.022	0.42 ± 0.024
Z_3	**0.59 ± 0.031**	**0.73 ± 0.044**	**0.63 ± 0.032**	**0.85 ± 0.049**	**1.18 ± 0.056**	**1.43 ± 0.84**	**0.45 ± 0.029**	**0.64 ± 0.037**
Z_4	**0.62 ± 0.036**	**0.92 ± 0.058**	**0.77 ± 0.048**	**0.64 ± 0.038**	**0.58 ± 0.036**	**0.87 ± 0.053**	**0.29 ± 0.018**	**0.34 ± 0.021**

Table 3.7 Accuracy of the method of polarization-phase reconstruction of optical anisotropy maps of samples of biopsy of connective tissue component of the ligament of the uterus

Parameters	Z_i	$\Delta n_{0;90}$ (%)	$\Delta n_{\otimes;\oplus}$ (%)	$\Delta\mu_{0;90}$ (%)	$\Delta\mu_{\otimes;\oplus}$ (%)
$Ac(Z_i)$	Z_1	76.2	72.3	82.4	68.3
	Z_2	80.4	76.2	84.3	72.4
	Z_3	**88.1**	**87.4**	**90.3**	**88.2**
	Z_4	**90.4**	**88.2**	**90.2**	**87.3**

- a good level of balanced accuracy $\left(Ac\left(Z_{3;4}\left(\Delta n_{0;90}; \Delta n_{\otimes;\oplus}; \Delta\mu_{0;90}; \Delta\mu_{\otimes;\oplus}\right)\right)\right) \sim$ 90%) of the method in diagnosing the pathology of this organ. (depicted by bold —Table 3.7).

3.3 Differential Mapping of the Diffuse Component of the Polarization Matrix of Myosin Networks of Myocardium Samples

This part of the paper contains materials for studying the distribution of the magnitudes of optical anisotropy fluctuations of spatially structured myosin networks of myocardial layers by mapping the distributions of the magnitude of the elements of a second-order differential matrix. An analysis of the data of the indicated diffuse tomography technique was carried out under the assumption of a "spectral" predominance of the mechanisms of phase anisotropy $\left\{ \begin{array}{l} \langle\delta\rangle > \langle\tau\rangle; \\ \langle\varphi\rangle > \langle\chi\rangle \end{array} \right\}$. Based on this,

the main information parameters in a series of diffuse tomograms $\left\{ \begin{array}{l} \tilde{T}(\tilde{k}_{22}) \\ \tilde{T}(\tilde{k}_{33}) \\ \tilde{T}(\tilde{k}_{44}) \end{array} \right\}(x, y)$

for us are the coordinate distributions of the fluctuations of the parameters of linear $(\theta_4(F_{0;90})$ and $\theta_5(F_{45;135}))$ and circular birefringence $(\theta_6(F_{\otimes;\oplus}))$ (Table 3.8).

By analogy with the data of the polarization-phase reconstruction of the phase anisotropy of the polycrystalline structure of biological layers, we shall consider the parameters of fluctuations of linear (θ_δ) and circular (θ_φ) birefringence

$$\left\{ \begin{array}{l} \theta_\delta = \sqrt{\theta_4^2 + \theta_5^2} = \tilde{k}_{11}^{-1}\sqrt{\tilde{k}_{22}^2 + \tilde{k}_{33}^2}; \\ \theta_\varphi \equiv \theta_6 = \tilde{k}_{11}^{-1}\tilde{k}_{44}. \end{array} \right. \tag{3.1}$$

Table 3.8 Statistical moments of the 1st—4th orders that characterize distributions $T(x, y)$ of histological sections of the myocardium

$Z_{i=1;2;3;4}$	$T(\delta)$		$T(\varphi)$		$T(\tau)$		$T(\chi)$	
	Type A	Type B	Type A	Type B	Type A	Type B	Type A	Type B
$Z_1 \times 10^{-1}$	0.12 ± 0.008	0.072 ± 0.0045	0.084 ± 0.0056	0.077 ± 0.0045	0.08 ± 0.0067	0.045 ± 0.0044	0.074 ± 0.0045	0.068 ± 0.0054
$Z_2 \times 10^{-1}$	0.16 ± 0.012	0.089 ± 0.0076	0.14 ± 0.012	0.12 ± 0.0082	0.11 ± 0.0078	0.09 ± 0.0055	0.121 ± 0.0089	0.098 ± 0.0075
Z_3	$\mathbf{0.88 \pm 0.061}$	$\mathbf{0.52 \pm 0.038}$	1.09 ± 0.087	0.87 ± 0.063	$\mathbf{1.26 \pm 0.091}$	$\mathbf{0.93 \pm 0.067}$	1.35 ± 0.107	1.16 ± 0.092
Z_4	$\mathbf{1.15 \pm 0.093}$	$\mathbf{0.69 \pm 0.044}$	$\mathbf{1.45 \pm 0.11}$	$\mathbf{1.15 \pm 0.095}$	$\mathbf{1.85 \pm 0.16}$	$\mathbf{1.08 \pm 0.091}$	1.57 ± 0.15	1.36 ± 0.14

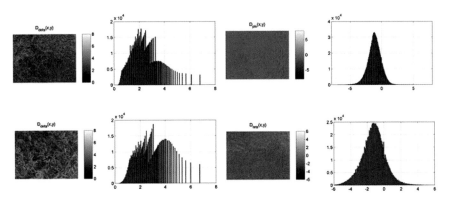

Fig. 3.7 Maps ((1), (3), (5), (7)) and histograms ((2), (4), (6), (8)) of the distributions fluctuation values of linear birefringence θ_δ and dichroism θ_φ of myocardium fibrillar networks layers of the dead patients from type A ((1)–(4)) and type B ((5)–(8))

On the fragments of Fig. 3.7 shows the results (maps and histograms) of the method of diffuse Mueller-matrix tomography of the fluctuations of linear $(\tilde{T}(\theta_\delta))$ and circular $(\tilde{T}(\theta_\varphi))$ birefringence.

An analysis of the coordinate distributions of diffuse parameters $(\theta_\delta(x, y), \theta_\varphi(x, y))$ of optical anisotropy implies:

- for myocardial samples with severe degenerative-dystrophic changes (type B), the fluctuations of linear birefringence θ_δ and dichroism θ_ϕ of structured myosin networks are smaller in comparison with similar distributions $N(\theta_\varphi)$, $N(\theta_\varphi)$ obtained for myocardium type A samples (Fig. 3.7, (1)–(8))
- Statistically this scenario illustrates the increase in the values of statistical moments of the 1st and 2nd orders that characterize the histograms $N(\theta_\varphi)$ and $N(\theta_\varphi)$ (Fig. 3.7, (6), (8)).

3.4 Differentiation of Polarization-Phase Diffuse Maps of Optical Anisotropy of Biological Preparations of Various Morphological Structures and Pathological Conditions

The data obtained using the methods of Mueller-matrix differential polarization-phase and diffuse tomography (type A and type B myocardium; connective tissue component of the vaginal wall type C and type D) were the basis for a comparative statistical analysis of ensembles of average values and the magnitude of fluctuations of linear birefringence and dichroism.

Table 3.9 Strength parameters of the polarization reconstruction method of the phase and amplitude anisotropy parameters of myocardium layers

$Ac(Z_i)$	$Z_{i=1;2;3;4}$	$T(\delta)$ (%)	$T(\varphi)$ (%)	$T(\tau)$ (%)	$T(\chi)$ (%)
	Z_1	86.5	62.4	74.4	61.3
	Z_2	88.4	65.2	68.2	65.3
	Z_3	**91.2**	76.3	78.5	72.4
	Z_4	**95.4**	78.5	80.4	69.5

Table 3.10 Statistical moments of the 1st–4th orders that characterize distributions of diffuse tomograms of linear birefringence $\tilde{T}(\theta_\delta)$ and dichroism $\tilde{T}(\theta_\varphi)$ of histological sections of the myocardium

$Z_{i=1;2;3;4}$	$\tilde{T}(\theta_\delta)$		$\tilde{T}(\theta_\varphi)$	
	Type A	Type B	Type A	Type B
$Z_1 \times 10^{-1}$	**0.32 ± 0.017**	**0.43 ± 0.024**	**0.094 ± 0.0056**	**0.12 ± 0.0067**
$Z_2 \times 10^{-1}$	**0.22 ± 0.013**	**0.35 ± 0.017**	**0.088 ± 0.0045**	**0.093 ± 0.0077**
Z_3	1.23 ± 0.13	0.92 ± 0.12	0.83 ± 0.067	0.68 ± 0.034
Z_4	0.98 ± 0.062	0.74 ± 0.041	1.34 ± 0.14	1.04 ± 0.11

The method of differential diagnostics of statistical and informational data analysis of the distribution of the magnitude of statistical moments of the 1st–4th orders, which characterize the distribution of the fluctuations of the linear birefringence and dichroism, is presented in detail in Sect. 3.2.2 of our work.

Established (Table 3.9) the excellent accuracy $(Ac(Z_{3;4}(\delta)) > 90\%)$ of differential diagnosis of the severity of degenerative-dystrophic changes in optically anisotropic myocardial networks by calculating the asymmetry and excess of polarized-reconstructed distributions of linear birefringence. The strength of the matrix polarization tomography method based on a statistical analysis of the distributions of circular birefringence (φ) and optically anisotropic absorption $(\tau; \chi)$ of myocardial samples does not exceed the satisfactory level—$Ac(Z_{i=1-4}) < 80\%$.

Diagnostic capabilities of the matrix polarization reconstruction technique for maps of parameter fluctuations of phase anisotropy of myosin networks of myocardial samples are presented in Table 3.10.

The statistical moments most sensitive to changes in the distributions of the diffuse parameters of the optical anisotropy of the myocardium—average and dispersion (highlighted in bold). Within both groups of histological preparations of the myocardium, the following difference levels are realized—Z_1 (1.42 times) and Z_2 (1.61 times).

An information analysis of the results (Table 3.11) revealed the following levels of balanced accuracy of the matrix reconstruction method of the diffuse component of the polycrystalline structure of the A and B types of myocardium depolarizing layers.

Table 3.11 Strength parameters of the polarization reconstruction method for phase anisotropy fluctuations of myocardium layers

$Ac(Z_i)$	$Z_{i=1;2;3;4}$	$\tilde{T}(\theta_\delta)$ (%)	$\tilde{T}(\theta_\varphi)$ (%)
	Z_1	**83.4**	66.2
	Z_2	**85.5**	68.4
	Z_3	75.3	62.1
	Z_4	77.4	60.3

Table 3.12 Strength parameters of the method of polarization reconstruction of average values and magnitude of fluctuations in phase anisotropy of the vaginal wall

$Ac(Z_i)$	$Z_{i=1;2;3;4}$	$T(\delta)$ (%)	$T(\tau)$ (%)	$\tilde{T}(\theta_\delta)$ (%)	$\tilde{T}(\theta_\varphi)$ (%)
	Z_1	81.2	73.3	**87.4**	80.5
	Z_2	82.4	68.2	84.2	**88.3**
	Z_3	**87.3**	**85.4**	79.3	66.4
	Z_4	**91.5**	82.1	**85.5**	64.2

The results obtained (Table 3.11) illustrate a good level of accuracy in the differentiation of histological sections of the A and B types of myocardium —$Ac\big(Z_{1;2}(\theta_\delta)\big) \sim 83 - 85\%$.

A similar comparative analysis of the statistical reliability and informational strength of the methods of matrix polarization reconstruction of the distributions of average values and fluctuations of the optical anisotropy parameters of biological crystallites of healthy tissue of the vaginal wall and of that with genitals prolapse is shown in Table 3.12.

As in the expert evaluation of degenerative-dystrophic changes in myocardial tissue, the accuracy of the clinical application of this methods of polarization-phase and diffuse Mueller matrix tomography in differential diagnosis of the connective tissue component of the vaginal wall at genital prolapse reaches the good (85–90%) and high (>90%) quality (highlighted in bold).

Note that the considered methods of polarization correlometry and differential Mueller-matrix mapping have effectively proved themselves in the differential diagnosis of changes in the polycrystalline structure of partially depolarizing layers of biological tissues. The urgent task follows from this—determining the sensitivity of the matrix and polarization-correlation techniques for diagnosing optical anisotropy of strongly depolarizing diffuse layers of biological tissues.

3.5   Structural Analysis of the Coordinate Distributions of the Elements of the Mueller Matrix of Depolarizing Layers Biological Tissues

At the first stage, within the framework of mutually complementary statistical, autocorrelation, and scale-self-similar approaches, we investigated the structure of Mueller-matrix images of model and real strongly scattering objects.

For the experiments, we used two groups of samples: (i) tissue phantoms; (ii) paraffin blocks of excised normal and cancerous colorectal tissues in vitro.

(i) The studied tissue phantoms represent an assembly of rutile (TiO_2) particles (mean size—0.53 µm, standard deviation—0.01 µm, electron microscopy measurements) which are embedded homogeneously in a PVC-based host matrix. Each sample was prepared in the form of 1 mm thick 2 cm plastic sheet using different concentrations of TiO_2 particles, namely, 1.5, 3.0, and 6.0 mg/ml. This ensured a controlled experimental implementation (due to variation of scattering coefficients μ_s) of the light scattering frequency, from a nondepolarizing layer with a single interaction to a strongly scattering (depolarizing) phantom.

(ii) The paraffin blocks of excised colorectal human tissue were prepared according to conventional protocol for pathology analysis.

The experimental Mueller-polarimetric measurements of diffuse layers were carried out in two modes:

• transmission—a series of phantoms with different light scattering;
• reflection—the tissue paraffin blocks.

The results of measurement and analysis (within the statistical, correlation, and fractal approaches) of the coordinate distributions of the degree of depolarization of phantoms with different scattering powers for wavelength $\lambda = 550$ nm are presented in a series of fragments Fig. 3.8

Here:

• coordinate distribution of depolarization Δ (Fig. 3.9, (1)–(3));
• histograms $N(\Delta)$ of the distribution of depolarization magnitude Δ (Fig. 3.9, (4)–(6));
• autocorrelation functions $AK_\Delta(\Delta x)$ of the distribution of depolarization magnitude Δ (Fig. 3.9, (7)–(9));
• the logarithmic dependencies of power spectra $\log S(\Delta) - \log l^{-1}$ of the distribution of depolarization magnitude Δ (Fig. 3.9, (10)–(12));
• a series of phantoms of various optical thicknesses $\mu_S = 2.5\,\text{mm}^{-1}$ (Fig. 3.9, (1), (4), (7), (10)); $\mu_S = 5\,\text{mm}^{-1}$ (Fig. 3.9, (2), (5), (8), (11)); $\mu_S = 10\,\text{mm}^{-1}$ (Fig. 3.9, (3), (6), (9), (12)).

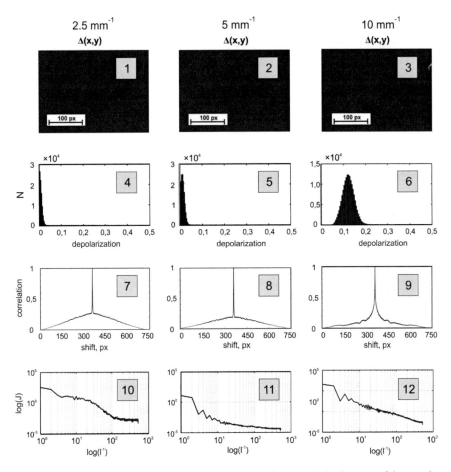

Fig. 3.8 Statistical, correlation and fractal characteristics of the depolarization map of the samples 1–3

An analysis of the results (Fig. 3.8) showed:

- a shift towards large values of the extremum of the histograms of the distributions of depolarization of phantoms with increasing of scattering coefficient from 2.5 to 10 mm^{-1} (Fig. 3.9, (4)–(6));
- increasing the range of variation of random values of Δ with the increase of mean value of depolarization (Fig. 3.9, (4)–(6));
- the different phantom samples are characterized by the "individual" half-width, asymmetry and sharpness of the distributions $N(\Delta)$ peak;
- coordinate heterogeneity of distributions depolarization $\Delta(p \times q)$ of the scattering tissue phantoms—an increase in the decay rate of the eigenvalues of the series of the autocorrelation functions A $K_{\Delta}(\Delta x)$ (Fig. 3.9, (7), (8));

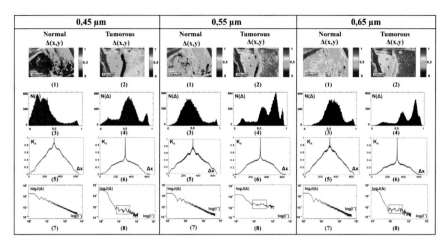

Fig. 3.9 Depolarization maps Δ—(1, 2), histograms $N(\Delta)$—(3, 4), autocorrelation functions $AK_\Delta(\Delta x)$—(5, 6) and power spectra (log $S(\Delta)$—log l^{-1})—(7, 8) presented for the samples with normal tissue and colorectal carcinoma, respectively, obtained for the measurement wavelengths: $\lambda_1 = 450$ nm, $\lambda_2 = 550$ nm, $\lambda_3 = 650$ nm. Explanations are given in the text

Table 3.13 Quantitative statistical (Z_i), autocorrelation (K_j), and scale-self-similar (D^f) parameters that characterize distributions $\Delta(p \times q)$ of phantom samples 1–3

	λ (µm)	Z_1	Z_2	Z_3	Z_4	K_2	K_4	D^f
Sample 1	0.53	0.017	0.0043	0.41	3.03	0.094	6.83	0.025
Sample 2	0.52	0.014	0.0087	0.34	3.08	0.27	2.34	0.052
Sample 3	0.55	0.16	0.15	0.19	0.19	0.33	0.43	0.18

- the formation of a large-scale self-similar (fractal) structure of depolarization maps with an increase in optical thickness—for the samples 1 and 2, the logarithmic dependencies log $S(\Delta)$ − log l^{-1} do not have a constant slope angle η of the approximating curves $\Phi(\eta)$ (Fig. 3.9, (10), (11)). Distribution $\Delta(p \times q)$ for the sample 3 is multi-fractal (two values of η) (Fig. 3.9, (12)).

The results of the statistical, correlation and fractal analysis of $\Delta(p \times q)$ for the wavelength 550 nm are shown in Table 3.13.

A comparative analysis of quantitative statistical, correlation, and fractal parameters of coordinate distributions of the magnitude of depolarization by phantoms with different light scattering ratios (from 2.5 to 10 mm^{-1}) revealed:

- with increasing optical thickness increases to 24 times the magnitude of the statistical moment of the second order, which characterizes the distributions $\Delta(p \times q)$;

– an increase (20 times) in the magnitude of the excess, which characterizes the sharpness of the peak of the distribution of the magnitude of depolarization by phantoms with a greater light scattering ratio;

– a 16-fold increase in fourth-order correlation moment, which characterizes the sharpness of the autocorrelation functions $AK_\Delta(\Delta x)$ peak with increasing scattering coefficient by phantom samples;

– growth (7 times) of the dispersion D^f magnitude of the extrema of logarithmic dependencies $\log S(\Delta) - \log l^{-1}$ of the distribution of the magnitude of the depolarization of phantoms with a greater light scattering.

With an increase in the depolarizing ability of phantom samples (scattering coefficient from 2.5 to 10 mm^{-1}), the basic laws of transformation of the values of statistical, correlation and fractal parameters are determined: $Z_1 \uparrow; Z_2 \uparrow; Z_3 \downarrow; Z_4 \downarrow;$ $K_2 \uparrow; K_4 \downarrow; D^f \uparrow$.

Thus, the study of a complex of objective parameters that characterize the statistical, coordinate-inhomogeneous, and scale-like structure of depolarization maps of a series of model phantom layers showed a high sensitivity of statistical (Z_i), correlation (K_j), and fractal (D^f) moments to changes in light scattering [4–6]. The results obtained demonstrated the promise of using depolarization maps in the differential diagnosis of real strongly scattering layers of biological tissues.

The results of this diagnosis are presented in a series of fragments Fig. 3.9 for the paraffin blocks of normal (panels 1, 3, 5, 7) and cancerous (panels 2, 4, 6, 8) colorectal tissue.

Here:

• coordinate distributions of the degree of depolarization Δ [7] (panels 1, 2);
• statistical distributions of the degree of depolarization Δ—histograms $N(\Delta)$ (panels 3, 4);
• correlation distributions the degree of depolarization Δ—autocorrelation functions $AK_\Delta(\Delta x)$ (panels 5, 6);
• fractal distributions the degree of depolarization Δ—the logarithmic dependencies of the power spectra $\log J(\Delta) - \log l^{-1}$(panels 7, 8)

The obtained data (see Fig. 3.9a–c) display an explicit spectral dependence of the group of statistical (Fig. 3.9, panels 3, 4), 2l (Fig. 3.9, panels 5, 6) and fractal $\log S(\Delta) - \log l^{-1}$ (Fig. 3.9, panels 7, 8) parameters which characterize the distribution of the degree of depolarization (Fig. 3.9, panels 1, 2) measured in reflection on flat surfaces of paraffin blocks with normal and cancerous tissues used for histological sectioning.

The peaks of distributions of degree of depolarization Δ for such samples are located within the range $0, 25 \leq \Delta \leq 0.85$(Fig. 3.9, panels 3, 4). The comparison of Δ value distributions for the phantom samples (Fig. 3.9, (4)–(6)) and biological tissue reveals that latter distributions are multi-modal with several peaks. With wavelength increase the peaks of parameter Δ histograms are shifted to larger values for both healthy and cancerous samples (Fig. 3.9, panels 3, 4). Besides, the range of Δ values expands. This fact indicates the 'longwave' increase of depolarization of the radiation rearranged by the biological layers. We attribute this fact to deeper penetration of long wavelengths into the tissue. Longer optical path results in larger number of scattering events which randomize the polarization of incident light. Furthermore, the depolarization degree of cancerous tissue is higher than that of the normal tissue sample. It is worth to mention that an opposite trend on depolarization was observed in the experiments with fresh thick tissue specimens (colon, uterine cervix) when epithelial surface of tissue was imaged. In the above mentioned measurement configuration an imaging plane was orthogonal to the plane of tissue histological cuts seen and analyzed by pathologist.

Two-dimensional distributions of depolarization $\Delta(p \times q)$ have complex coordinate-heterogeneous topographic structure (Fig. 3.9, panels 1, 2) in comparison with similar maps for the phantom samples (Fig. 3.9, (1)–(3)). Obviously, this difference is related to the morphological structure of biological tissues. Quantitatively, the topographic heterogeneity of depolarization maps $\Delta(p \times q)$ is illustrated by the trend of autocorrelation functions $AK_\Delta(\Delta x)$ of the healthy and cancerous tissue (Fig. 3.9, panels 5, 6), which have a larger half-width $K_2 \uparrow$ and less sharp peak $K_4 \downarrow$ as compared to the phantom samples due to a larger range of values of the degree of depolarization (Tables 3.14 and 3.15).

The analysis of obtained data revealed the most sensitive parameters in differentiation of normal and cancerous tissue $q(\lambda) \equiv \{ Z_1(\lambda); \ Z_3(\lambda); \ Z_4(\lambda); K_4(\lambda); \ D(\lambda) \}$:

Table 3.14 Statistical (Z_i), correlation (K_j) and fractal (D^f) parameters, which characterize the distributions $\Delta(p \times q)$ of samples of real biological tissues

λ	0.45 µm		0.55 µm		0.65 µm	
Samples	Normal	Tumor	Normal	Tumor	Normal	Tumor
Z_1	0.23	0.51	0.34	0.77	0.43	0.82
Z_2	0.28	0.33	0.35	0.38	0.39	0.36
Z_3	1.24	0.76	0.68	0.93	0.45	1.76
Z_4	1.03	0.65	0.45	1.97	0.32	1.63
K_2	0.21	0.35	0.32	0.37	0.33	0.41
K_4	0.74	0.58	0.49	0.28	0.36	0.14
D^f	0.14	0.18	0.17	0.28	0.14	0.32

Table 3.15 The statistical structure parameters of layered diffuse tomograms of the fluctuation phase anisotropy of a polycrystalline blood film from healthy donors

DT_{red}	$DT_{red}(\tilde{BL}_{0:90})$			$DT_{red}(\tilde{BL}_{45:135})$			$DT_{red}(\tilde{BC}_{\otimes:\oplus})$		
$\varphi_j\,\text{(rad)}$	0.6	0.9	1.2	0.6	0.9	1.2	0.6	0.9	1.2
X_1	0.089	0.14	0.15	0.053	0.094	0.13	0.17	0.14	0.088
X_2	0.078	0.11	0.14	0.044	0.08	0.11	0.14	0.12	0.084
X_3	0.35*	0.27*	0.22*	0.65*	0.54*	0.33*	0.32*	0.18*	0.12*
X_4	0.31*	0.23*	0.19*	0.91*	0.77*	0.45*	0.39*	0.32*	0.23*

- 1st group—(non-colored fragments)—the differences between parameters do not exceed 25–45%;
- 2nd group—(highlighted in yellow)—the differences between parameters are within 1.5–2 times;
- 3rd group—(highlighted in italic)—the differences between parameters are within 2–5 times.

Spectral analysis shows that with the increase of probe beam wavelength, coordinate uniformity of the distributions $\Delta(p \times q)$ grows for both healthy and cancerous samples due to the depolarization enhancement. From the quantitative point of view, this trend is illustrated by the increase of the FWHM and decrease of the peak sharpness of the dependencies $AK_\downarrow(\Delta x))$ (Fig. 3.9, panels 5, 6). The most explicit case of this scenario is illustrated by the depolarization maps of the cancerous tissue (Fig. 3.9, panels 5, 6). As it can be seen from Table 3.1, along with wavelength increase $(\lambda \uparrow)$ the parameter K_2 increases and the parameter K_4 decreases. Comparative analysis of the logarithmic dependencies $\log S(\Delta) - \log l^{-1}$ calculated for the distributions $\Delta(p \times q)$ of biological samples, reveals significant differences for normal and cancerous tissues. It has been shown, that the distributions of depolarization $\Delta(p \times q)$ for normal tissue are fractal in all spectral ranges (Fig. 3.9, panels (7)). For cancerous sample, the distributions $\Delta(m \times n)$ are random in the region of medium sizes (~ 100–300 μm) (Fig. 3.9, panels (8), rectangular boxes). The discovered pattern is in a good correlation with the data obtained by Mueller-matrix mapping of optically-thin histological cuts of different organ tissues (prostate, cervix and uterine wall) where the 'oncological destruction' of fractality of two-dimensional structure of depolarization maps was associated with the formation of new fibrillar networks.

3.6   3D Matrix Reconstruction of the Polycrystalline Structure of Depolarizing Blood Films for Cancer Diagnosis

Nowadays, Mueller-matrix polarimetry (MMP) approaches are extensively used for the visualization of malformation of biological tissue structure and determination of its functional physiological variations. The MMP now includes a number of popular research directions, such as studies of scattering matrices, polar decomposition of Mueller matrices, two-dimensional Mueller-matrix mapping, and others. The main disadvantage of MMP is associated with invasive procedure of preparation of bio-tissue samples that significantly limits its application for biomedicine. In fact, the easily accessible biological liquids, including those obtained from the particular organs, could be used as the main subjects of tissue samples in MMP. Thus, one of the most promising applications of MMP is screening of the so-called 'integral' fluid of the human body, such as blood and plasma. The film of such biological fluids represents a complex spatially inhomogeneous optically anisotropic structure, which is formed by various types of biochemical and molecular crystalline complexes. Monitoring of the dynamics of crystallization within such thin films provides an opportunity to characterize the internal processes at the macro-level of molecular interaction and to carry out the early diagnosis of various diseases.

Here, we examine the polarimetric diagnosis towards the possibility of quantitative characterization of the optical anisotropy of polycrystalline films of whole blood. This object of investigation can be considered both as a heterogeneous complex-structured liquid and/or heterogeneous colloidal polymer solution. Traditionally, the molecular components (including various proteins—albumin monomers ($\sim 60\%$), globulin ($\sim 40\%$), and fibrinogen) and the blood cells (erythrocytes, platelets, and leukocytes) are investigated by biochemical methods. At normal conditions, the protein fraction is typically presented in the form of monomers, whereas at the oncological pathology, the tertiary and quaternary structure of proteins are changed with the invariance of biochemical parameters. A few quite effective methods are available and used for studying supramolecular (dimers, trimers, and monomers) protein structures (100–2000 nm) of blood. In fact, a development of new and more sensitive and informative screening technologies for diagnosis of structural protein changes in tissue samples are required. The polarimetry of polycrystalline films of biological fluids is one of extremely promising technique for this role.

The feasibility studies of polarimetric examinations of optically thin non-depolarizing polycrystalline plasma films have been presented. In the frame of these studies, a development of algorithms for reconstructing the distributions of phase and amplitude anisotropy parameters of the protein polycrystalline networks of sampling layers have been performed.

In practical use, the MMP of blood and plasma samples is associated with two main limitations. The first one is that most of polycrystalline plasma films are partially depolarized incident light, which leads to a decrease of accuracy of differential diagnosis. The second one is that the biochemical composition of plasma is not so informative, as compared to blood itself. At the same time, due to spatially inhomogeneous structure and the presence of uniform elements, the blood samples provide a higher level of depolarization.

These factors complicate the direct reconstruction of the optical anisotropy distributions in the samples with subsequent differentiation of its morphological changes. Therefore, to develop a new generalized MMP approach for examination of multi-layered polycrystalline structures, such as diffuse biological tissue samples, are urgently required. This goal can be achieved by the combination (synthesis) of the two following techniques:

- Isolation of the depolarized component of Mueller matrix of thin blood films by its decomposition on the basis of differential matrices of first order (i.e., the polarized part, which is the distribution of the mean values of the optical anisotropy parameters of polycrystalline structures of proteins and formed elements) and the second one (i.e., the depolarized part, defined by distribution of fluctuations of linear and circular birefringence and dichroism),—раздел 1.3, соотношения (1.1)–(1.30);

- The use of a coherent reference wave and the algorithm for digital holographic reconstruction of the complex amplitudes field in different sections of the blood film (раздел 1.5, соотношения (1.56)–(1.62)).

This part of our work is aimed to developing and experimental testing of the Mueller-matrix diffuse tomography approach for the layer-by-layer reconstruction of fluctuations of optical anisotropy within the thin tissue samples, with the final target to differentiate polycrystalline blood films of healthy donors and patients with prostate cancer.

3.6.1 Samples

Blood from 18 healthy donors (group 1) and 18 patients with prostate cancer (group 2) were used in this study. Each sample film has been obtained by applying a drop of biological fluid to a substrate of optically homogeneous glass, followed by drying at room temperature. The optical thickness (τ) of the sample films varied within the range $0.68 \leq \tau \leq 0.75$, the degree of depolarization Δ—$41\% \leq \Delta \leq 52\%$.

Firstly, the functional possibility of optical anisotropy parameters 3D reproduction has been considered using the example of a polycrystalline blood film from healthy donor. Further, possibility of differentiation between polycrystalline whole blood films taken from healthy donors and patients with prostate cancer has been investigated.

3.6.2 Results and Discussion

3.6.2.1 Layered Maps of Fluctuations in the Parameters of Phase Anisotropy of a Partially Depolarizing Polycrystalline Film of Blood

Figure 3.10 presents the diffuse tomograms $\mathbf{DT_{red}}\left(\tilde{\mathbf{BL}}_{0;90}\right)$ ((1)–(3)), $\mathbf{DT_{red}}$ $\left(\tilde{\mathbf{BL}}_{45;135}\right)$ (4)–(6), $\mathbf{DT_{red}}\left(\tilde{\mathbf{BC}}_{\otimes;\oplus}\right)$ (7)–(9)) in the phase sections $\varphi_1 = 0.6\,\mathrm{rad}$ ((1), (4), (7)), $\varphi_2 = 0.9\,\mathrm{rad}$ (2), (5), (8)), $\varphi_1 = 1.2\,\mathrm{rad}$ (3), (6), and (9)) fluctuations of linear and circular birefringence of a partially depolarizing polycrystalline blood film from healthy donor ($\tau = 0.71; \Delta = 46\%$).

Analysis of the diffuse tomograms $\mathbf{DT_{red}}\left(\varphi_j, x, y\right)$ of an optically anisotropic blood film, presented at Fig. 3.10, revealed a good correlation between the experimental data and theoretical data:

- Individuality of layer wise coordinate distributions of linear $\left(\left(\tilde{\mathbf{BL}}_{0;90}\right);\right.$ $\left.\left(\tilde{\mathbf{BL}}_{45;135}\right)\right)$ and circular $\left(\left(\tilde{\mathbf{BC}}_{\otimes;\oplus}\right)\right)$ birefringence parameters (Fig. 3.10) fluctuations;

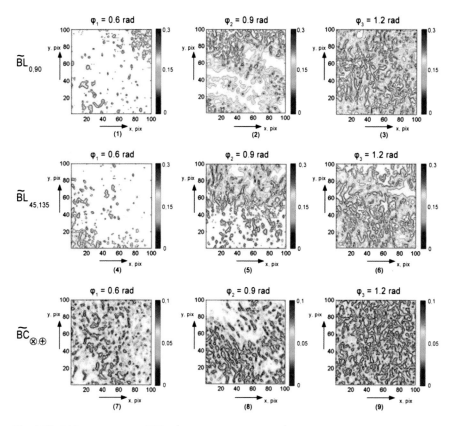

Fig. 3.10 Diffuse tomograms $\mathbf{DT}_{\mathbf{red}}\left(\varphi_j; 1000\,\mu m \times 1000\,\mu m\right)$ of linear and circular birefringence fluctuations of a polycrystalline blood film from practically healthy donors ($\tau = 0.71; \Delta = 46\%$). Explanations are in the text

- Dependence of the structure $\mathbf{DT}_{\mathbf{red}}\left(\varphi_j; x; y\right)$ on the value of the phase section φ_j;

- Increasing ($\uparrow$) of amplitude fluctuations $\mathbf{DT}_{\mathbf{red}} \left\{ \begin{array}{l} \left(\tilde{\mathbf{BL}}_{0;90}\right); \\ \left(\tilde{\mathbf{BL}}_{45;135}\right); \\ \left(\tilde{\mathbf{BC}}_{\otimes;\oplus}\right) \end{array} \right\} (x, y)$ with

growing of $\varphi_j \uparrow$.

In each phase cross-section, the coordinate distributions $\mathbf{DT}_{\mathbf{red}}\left(\varphi_j; x; y\right)$ were estimated by calculating the aggregate of statistical moments of the first–fourth order $X_{i=1;2;3;4}$.

Table 3.15 presents a series of "phase" dependences of the value $X_{i=1;2;3;4}$ that characterize the distributions $\left\{ \begin{array}{l} \mathbf{DT_{red}}\left(\mathbf{B\tilde{L}}_{0;90}\right); \\ \mathbf{DT_{red}}\left(\mathbf{B\tilde{L}}_{45;135}\right); \\ \mathbf{DT_{red}}\left(\mathbf{B\tilde{C}}_{\otimes;\oplus}\right) \end{array} \right\}(x;\mathbf{y})$ of the polycrystalline blood film from healthy donors.

Analysis of the data presented in Table 3.15 revealed the following statistical scenario for the variation of the distribution of fluctuations in the parameters of optical anisotropy:

$$\varphi_j \uparrow \Leftrightarrow \left\{ \begin{array}{l} X_{1;2}\left(\mathbf{DT_{red}}\left(\mathbf{B\tilde{L}}_{0;90}; \mathbf{B\tilde{L}}_{45;135}; \mathbf{B\tilde{C}}_{\otimes;\oplus}\right)\right) \uparrow; \\ X_{3;4}\left(\mathbf{DT_{red}}\left(\mathbf{B\tilde{L}}_{0;90}; \mathbf{B\tilde{L}}_{45;135}; \mathbf{B\tilde{C}}_{\otimes;\oplus}\right)\right) \downarrow. \end{array} \right. \tag{3.2}$$

Such changes in magnitude $X_{i=1;2;3;4}$ can be attributed to the change in the multiplicity of light scattering in the volume of a polycrystalline film of blood. In the region of small φ_j distributions $\mathbf{DT_{red}}\left(\mathbf{B\tilde{L}}_{0;90}\right); \mathbf{DT_{red}}\left(\mathbf{B\tilde{L}}_{45;135}\right); \mathbf{DT_{red}}$ $\left(\mathbf{B\tilde{C}}_{\otimes;\oplus}\right)$, they are asymmetric $(X_{3;4} > X_{1;2})$. Due to the increase in multiplicity $(\varphi_j \uparrow)$ according to the central boundary theorem, the structure of such distributions tends to be normal $(X_{1;2} \uparrow; X_{3;4} \to 0)$.

The most sensitive to changes in the polarization manifestations of fluctuations in the phase anisotropy parameters of the polycrystalline blood film were statistical moments of the third and fourth orders. The range of their variations reaches (marked by *):

- $\mathbf{DT_{red}}\left(\mathbf{B\tilde{L}}_{0;90}\right)$—1.77 to 1.81 times;
- $\mathbf{DT_{red}}\left(\mathbf{B\tilde{L}}_{45;135}\right)$—1.83 to 1.89 times;
- $\mathbf{DT_{red}}\left(\mathbf{B\tilde{C}}_{\otimes;\oplus}\right)$—1.81 to 2.82 times.

3.6.2.2   3D Mueller-Matrix Differentiation of Diffuse Polycrystalline Films of Blood

Optical technology for differential diagnosis of depolarizing polycrystalline whole blood films from healthy donors (group 1) and patients with prostate cancer (group 2) includes:

1. Determination in each group of samples of a series of "phase" layered images of 3D distributions $\mathbf{DT_{red}}\left\{\mathbf{B\tilde{L}}_{0;90}; \mathbf{B\mathtt{L}}_{45;135}; \mathbf{B\tilde{C}}_{\otimes;\oplus}\right\}(\varphi_1 = 0.3\,\text{rad}; 2\varphi_1; \ldots, 6\varphi_1)$.
2. Calculations for each "phase" section ϕ_j of the statistical moments of the first–fourth order $X_{i=1;2;3;4}\left\{\mathbf{DT_{red}}\left[\mathbf{B\tilde{L}}_{0;90}; \mathbf{B\mathtt{L}}_{45;135}; \mathbf{B\tilde{C}}_{\otimes;\oplus}\right](\varphi_j, x, y)\right\}$.

3. Definition of "phase" planes (ϕ^*), where the differences $\Delta X_i = X_i^{(group1)} - X_i^{(group2)}$ between the statistical moments are maximum—$(\Delta X_{i=1;2;3;4}^* \equiv \Delta X_{i=1;2;3;4}(\varphi^*) \to \max)$.

4. In the phase plane φ^*, the mean $\bar{\Delta X}_{i=1;2;3;4}^*$ and $\eta(\Delta X_i^*)$ error within the polycrystalline blood films from group 1 and group 2 are determined.

5. For the possible clinical use of this method for each of the statistical moments $X_{i=1;2;3;4}(\phi^*)$—sensitivity $(\mathbf{R} = \frac{a}{a+b}100\%)$ specificity $(\mathbf{Q} = \frac{c}{c+d}100\%)$; balanced accuracy $(\mathbf{Ac} = \frac{R+Q}{2})]$ is calculated, where a and b are both the number of correct and incorrect diagnoses within group 2; c and d—the same within group 1. [3]

Figures 3.13, 3.14 and 3.15 show the phase $(\varphi^* = 0.85\,\text{rad})$ cross sections $(\Delta X_{i=1;2;3;4}^* \equiv \Delta X_{i=1;2;3;4}(\varphi^*) \to \max)$ of 3D distributions of diffuse tomograms $\mathbf{DT_{red}}(\tilde{\mathbf{BL}}_{0;90})$ (see Fig. 3.11), $\mathbf{DT_{red}}(\tilde{\mathbf{BL}}_{45;135})$ (see Fig. 3.12), $\mathbf{DT_{red}}(\tilde{\mathbf{BC}}_{\otimes;\oplus})$ (see Fig. 3.13) of polycrystalline blood films from healthy donors (fragments (1) and (2)) and patients with prostate cancer (fragments (3) and (4)).

Table 3.16 shows the statistical analysis of the "phase" cross-sections $\varphi^* = 0.85\,\text{rad}$, as well as the level of balanced accuracy $\mathbf{Ac}, \%$.

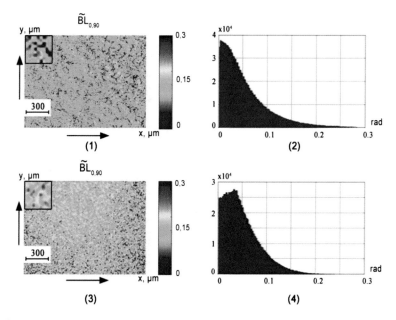

Fig. 3.11 Maps and histograms of the distributions of fluctuations $\mathbf{DT_{red}}$ of the linear birefringence $(\tilde{\mathbf{BL}}_{0\cdot90})$ of polycrystalline blood films from healthy donors ((**1**) and (**2**)) and cancer patients ((**3**) and (**4**))

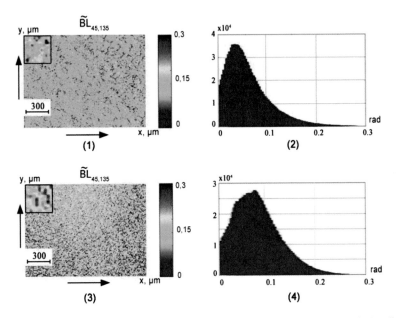

Fig. 3.12 Maps and histograms of the distributions of fluctuations DT_{red} of the linear birefringence $\left(\mathbf{B\tilde{L}}_{45;135}\right)$ of polycrystalline blood films from healthy donors ((**1**) and (**2**)) and cancer patients ((**3**) and (**4**))

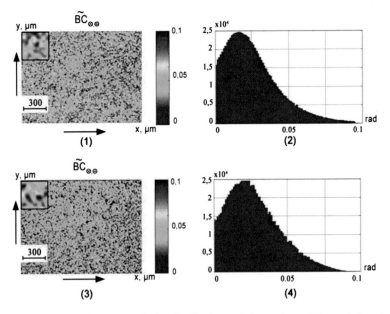

Fig. 3.13 Maps and histograms of the distributions of fluctuations DT_{red} of the circular birefringence $\left(\mathbf{B\tilde{L}}_{\otimes;\oplus}\right)$ of polycrystalline blood films from healthy donors ((**1**) and (**2**)) and cancer patients ((**3**) and (**4**))

Table 3.16 Efficiency of statistical analysis of diffuse tomograms of polycrystalline blood films for the differentiation between healthy donors and patients with prostate cancer patients

$\mathbf{DT_{red}}$	$\mathbf{B\tilde{L}_{0;90}}$		$\mathbf{Ac}\,(\%)$
Samples	Group 1	Group 2	
X_1	0.063 ± 0.0045	0.077 ± 0.0068	78.7
X_2	0.081 ± 0.0052	0.096 ± 0.0056	74.4
X_3	$0.48 \pm 0.023*$	$0.35 \pm 0.019*$	84.6
X_4	$0.37 \pm 0.018*$	$0.28 \pm 0.016*$	80.3
$\mathbf{DT_{red}}$	$\mathbf{B\tilde{L}_{45;135}}$		$\mathbf{Ac}\,(\%)$
Samples	Group 1	Group 2	
X_1	0.067 ± 0.0044	0.089 ± 0.0045	74.4
X_2	0.107 ± 0.0033	0.13 ± 0.0047	76.3
X_3	$0.75 \pm 0.042*$	$0.57 \pm 0.031*$	84.7
X_4	$1.04 \pm 0.054*$	$0.82 \pm 0.037*$	82.3
$\mathbf{DT_{red}}$	$\mathbf{B\tilde{C}_{\otimes;\oplus}}$		$\mathbf{Ac}\,(\%)$
Samples	Group 1	Group 2	
X_1	0.022 ± 0.0078	0.032 ± 0.012	82.4
X_2	0.031 ± 0.0052	0.043 ± 0.0074	78.3
X_3	$0.37 \pm 0.018*$	$0.22 \pm 0.013*$	92.6
X_4	$0.47 \pm 0.022*$	$0.34 \pm 0.018*$	90.7

A comparative analysis of the obtained data on the statistical structure of the coordinate distributions of the magnitude of fluctuations in the linear and circular birefringence of polycrystalline blood films revealed a good $(80\% \leq \mathbf{Ac}(X_{3;4}$ $(\mathbf{B\tilde{L}}_{0;90;45;135})) \leq 85\%)$ and excellent $(\mathbf{Ac}(X_{3;4}(\mathbf{B\tilde{C}}_{\otimes;\oplus})) \geq 90\%)$ level of balanced accuracy in the differential diagnosis of oncological prostate pathology (marked by * in Table 3.16).

These results can be explained by the fact that the prevailing blood from patients with prostate cancer is the mechanisms:

(1) The growth of linear birefringence fluctuations $(\mathbf{B\tilde{L}}_{0;90}; \mathbf{B\tilde{L}}_{45;135})$ due to the formation of supramolecular polycrystalline protein structures (albumin, fibrin);
(2) Increasing concentration and crystallization of birefringent leukocytes;
(3) The growth of circular birefringence fluctuations $(\mathbf{B\tilde{C}}_{\otimes;\oplus})$ due to the increase in the concentration and crystallization of optically active globulin molecules.

3.7   "Two-Point" Vector-Parametric Analysis of Polarization-Inhomogeneous Images of Networks of Biological Crystallites

This part of the work illustrates the applied capabilities and the effectiveness of differential diagnosis of another pathology of biological tissues— degenerative-dystrophic and necroic changes in the polycrystalline structure due to diabetes of various severity. С целью реализации данной задачи были:

- used biological tissue of the internal organs of rats;
- kidney tissue, a fibrillar tissue with a spatially structured birefringent network of collagen fibers, and liver tissue, parenchymatous tissue with optically active islands of protein complexes, were investigated;
- two representative samples of kidney and liver tissue samples (39 samples each) were formed — healthy and diabetic rats (21 days);
- non-depolarizing with single scatterings (thickness $l = 25\,\mu m \div 30\,\mu m$ attenuation coefficient $0.0093 \le \tau \le 0.0099$) native histological preparations were made;

We studied according to the technique presented in Sect. 3.2.1 (relations (1.56)–(1.60)) the polarization-correlation maps of phase-inhomogeneous object fields of selected native biological preparations of internal organs of rats.

The obtained distributions of the magnitude and phase of the "two-point" 3rd and 4th parameters of the Stokes vector were analyzed within the framework of complementary statistical, correlation, and scale-self-similar approaches.

In the series of Fig. 3.14, 3.15, 3.16 and 3.17 shows 2D distributions ((1), (5), statistical distributions (2), (6), autocorrelation functions 2D distributions (3), (7), logarithmic dependences of power spectra 2D distributions (4), (8)) of SCP modulus and SCP phase of microscopic images of samples of a healthy ((1), (2), (3), (4)) and pathologically changed ((5), (6), (7), (8)) rat kidney.

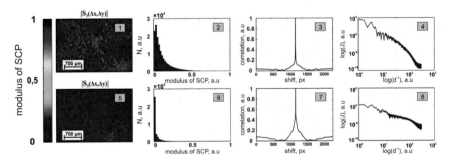

Fig. 3.14 Topographic (1), (5), statistical (2), (6), correlation (3), (7), fractal (4), (8) characteristics of SCP $|Sv_{i=3}(\Delta x; \Delta y)|$ of a healthy ((1), (2), (3), (4)) and pathologically changed ((5), (6), (7), (8)) rat kidney

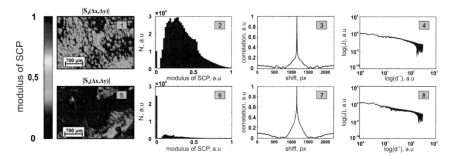

Fig. 3.15 Topographic (1), (5), statistical (2), (6), correlation (3), (7), fractal (4), (8) characteristics of SCP modulus $|Sv_{i=4}(\Delta x; \Delta y)|$ of a healthy ((1), (2), (3), (4)) and pathologically changed ((5), (6), (7), (8)) rat kidney

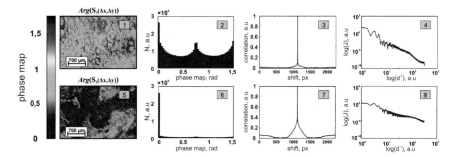

Fig. 3.16 Topographic (1), (5), statistical (2), (6), correlation (3), (7), fractal (4), (8) characteristics of SCP phase $Arg(Sv_{i=3}(\Delta x; \Delta y))$ of a healthy ((1), (2), (3), (4)) and pathologically changed ((5), (6), (7), (8)) rat kidney

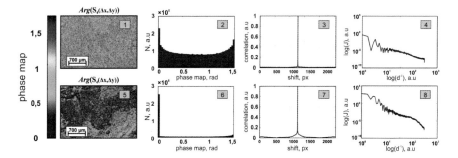

Fig. 3.17 Topographic (1), (5), statistical (2), (6), correlation (3), (7), fractal (4), (8) characteristics of SCP phase $Arg(Sv_{i=4}(\Delta x; \Delta y))$ of a healthy ((1), (2), (3), (4)) and pathologically changed ((5), (6), (7), (8)) rat kidney

3.7.1 Results and Discussion

3.7.1.1 Polarization-Correlation Maps of Samples of Rat Kidney

The basis of the analysis of the obtained data was the theoretical description (Sect. 1.4, relations (1.1)—(1.49)) of the relationships found between the analytical expressions of the "two-point" parameters of the Stokes vector of the polarization-inhomogeneous object field and the directions of the optical axes γ and phase shifts φ of the optically anisotropic biological layer.

The case of weak phase fluctuations is considered (Sect. 1.4, relations (1.50)–(1.55), Table 1.1). In this approximation, for the distributions of the modulus of the "two-point" parameters of the Stokes vector:

- It was established that for the polarization-correlation maps of the module of the 3rd and 4th "two-point" parameters of the Stokes vector, the most probable is the limiting value $|Sv_{i=3;4}(\Delta x, \Delta y)| = 1$ (see Figs. 3.14 and 3.15, fragments (1).
- It is physically justified that the values of the indicated polarization-correlation parameters of the object field of the birefringent biological layer are determined by the orientation structure of the directions of the optical axes of the fibrillar network.
- For a spatially ordered network (kidney tissue preparation) differences between the $\gamma(r_1)$ and $\gamma(r_2)$ are minor $(\gamma(r_1) - \gamma(r_2) \to 0)$ for the neighboring points (the scanning step $\Delta r = 1pix = 2\,\mu m$). Therefore, the following boundary values of modulus $|Sv_{i=3}(\Delta x, \Delta y)| \to 0$ and $|Sv_{i=3}(\Delta x, \Delta y)| \to 0$ are quite probable (see Figs. 3.14 and 3.15, fragments (2), (6)).
- The presence of other relationships between the directions of the optical axes of fibrils in a real optically anisotropic network of a kidney tissue preparation $\gamma(r_1) - \gamma(r_2) \neq 0$ is manifested in the formation of a series of local extrema values of SCP modulus in the entire range $0 \leq |Sv_{i=3;4}(\Delta x, \Delta y)| \leq 1$ (see Figs. 3.14 and 3.15, fragments (2), (6)).
- A comparison of the polarization-correlation maps of the modules of the 3rd and 4th parameters of the "two-point" Stokes vector of polarization-heterogeneous images of kidney tissue preparations from groups of healthy and diabetic rats was found the reduction of half-width of histograms $N^*(|Sv_{i=3;4}(\Delta x; \Delta y)|)$ of diabetic kidney (see Figs. 3.14 and 3.15 fragments (6)).

For coordinate distributions of the phase values of the 3rd and 4th parameters of the "two-pointed" Stokes vector of the boundary field of rat kidney tissue samples, it is shown:

- Differences in the distributions of local values of SCP phase maps of the 3rd and 4th parameters for spatially separated points of a polarization-inhomogeneous microscopic image (see Figs. 3.16 and 3.17, fragments (1), (5)).

- Differences between extreme phase values of the 3rd and 4th parameters of the "two-pointed" Stokes vector, which are determined by cross-correlation of parameters $\gamma(r)$ and the value of phase shift $\delta(r)$.
- Correspondence of the extreme phase values of the 3rd and 4th parameters of the "two-pointed" Stokes vector to theoretical analysis data —$Arg(Sv_3(\Delta x, \Delta y)) \to 0$ and $Arg(S_4(\Delta x, \Delta y)) \to 0.5\pi$, which are realized for a small scanning step of a polarization-inhomogeneous image

$$\Delta r = 2\,\mu m \to \begin{pmatrix} \gamma(r_1) \approx \gamma(r_2); \\ \delta(r_1) \approx \delta(r_2) \end{pmatrix}.$$

- The presence of major extrema in the statistical distributions of random phase variables of the 3rd and 4th parameters of the "two-pointed" Stokes vector— histograms are characterized by the corresponding main extrema $N(Arg(Sv_3(\Delta x, \Delta y)) = 0)$ and $N(Arg(Sv_4(\Delta x, \Delta y)) = 0.5\pi)$ for the full range of magnitude variation (see Figs. 3.16 and 3.17, fragments $0 \leq Arg(Sv_{i=3;4}(\Delta x, \Delta y)) \leq 0.5\pi$ (2), (6)).
- The decrease of the average and dispersion values characterizing the distributions of the SCP phase $Arg(Sv_{i=3;4}(\Delta x; \Delta y))$ of the object field of native preparations of the degenerative-dystrophically altered morphological structure of the fibrillar networks of the kidney tissue affected by diabetes is optically detected (see Figs. 3.16 and 3.17 fragments (2), (6)).
- The transformation of the topological structure of the polarization-correlation SCP-maps of the object field for the pathologically altered kidney samples is due to necrotic destruction of the structural anisotropy of fibrillar collagen networks. As a result, the coordinate and polarization inhomogeneities of the boundary field increase due to the fact that the small-scale optically anisotropic structure remains intact. Topographic changes in the 2D distributions of the values of SCP modulus and phase of polarization-structural microscopic images of experimental samples, which, as part of the statistical approach, are accompanied by the formation of an additional ensemble of values of $|Sv_{i=3}(\Delta x; \Delta y)| = 0$; $|Sv_{i=4}(\Delta x; \Delta y)| = 1$ and $Arg(Sv_{i=3}(\Delta x; \Delta y)) = 0$; $Arg(Sv_{i=4}(\Delta x; \Delta y)) = 0.5\pi$.
- The following scenario for changing the statistical structure (statistical moments of the 3rd and 4th orders) of polarization-correlation maps of the images of the sick kidneys sample $\begin{cases} Z_3\left(\left|Sv_{i=3;4}(\Delta x; \Delta y)\right|; Arg\left(Sv_{i=3;4}(\Delta x; \Delta y)\right)\right) \uparrow; \\ Z_4\left(Sv_{i=3;4}(\Delta x; \Delta y); Arg\left(Sv_{i=3;4}(\Delta x; \Delta y)\right)\right) \uparrow. \end{cases}$
- The decorrelation of the topographic distributions of SCP modulus and phase of phase-inhomogeneous object fields of samples of the pathological kidney changes the structure of the corresponding autocorrelation functions—the increase of half-width and the decrease in peak sharpness $\begin{cases} Z_2^K\left(Sv_{i=3;4}(\Delta x; \Delta y); Arg\left(Sv_{i=3;4}(\Delta x; \Delta y)\right)\right) \uparrow; \\ Z_4^K\left(Sv_{i=3;4}(\Delta x; \Delta y); Arg\left(Sv_{i=3;4}(\Delta x; \Delta y)\right)\right) \downarrow. \end{cases}$

- Fractal or multifractal self-similarity of the coordinate structure of SCP-maps ($|Sv_{i=3;4}(\Delta x, \Delta y)|$ and $Arg(Sv_{i=3;4}(\Delta x; \Delta y))$)—the least squares method revealed one or two slopes of the approximating curves of the logarithmic dependences $\log S(|Sv_{i=3;4}(\Delta x, \Delta y)|) - \log(d^{-1})$ and $\log S(Arg(Sv_{i=3;4}(\Delta x, \Delta y))) - \log(d^{-1})$ (see Figs. 3.14, 3.15, 3.16 and 3.17, fragments (4), (8)).
- A decrease in the value of the 2nd statistical moment D^f characterizing the dispersion $\log S(|Sv_{i=3;4}(\Delta x, \Delta y)|) - \log(d^{-1})$ and $\log S(Arg(Sv_{i=3;4}(\Delta x, \Delta y))) - \log(d^{-1})$ for polarization-correlation maps of the object field of native preparations of kidney tissue affected by diabetes, an optically anisotropic structure, which is more small scale in comparison with healthy tissue.

The diagnostic effectiveness of the statistical, correlation, and fractal methods of polarization-correlation (maps of the modules and phases of the 3rd and 4th parameters of the "two-point" Stokes vector) of differentiation of the two representative samples kidney samples illustrated by the data (average values and standard deviations) presented in Tables 3.17 and 3.18.

The results of quantitative processing of data about the statistical, correlation, and scale-self-similar structure of maps of the module of polarization-inhomogeneous images for the healthy and diabetic kidney samples :, shown in Table 3.17, revealed the following differences:

- Statistical moments of the 1st–4th orders—$\Delta Z_1 = 1.67$–8.11 times; $\Delta Z_2 = 1.85$–2.83 times; $\Delta Z_3 = 2.8$–3 times; $\Delta Z_4 = 2.21$–4.81 times;
- Correlation moments of the 2nd and 4th orders—$\Delta Z_2^k = 1.5$–1.57 times; $\Delta Z_4^k = 1.87$–2.53 times;
- Dispersion of the logarithmic dependences of the power spectra of the distributions of the magnitude modulus of the 3rd and 4th parameters of the "two-point" Stokes vector of object fields $\Delta D^f = 1.29$–1.52 times.

Quantitative statistical, correlation and fractal processing of phase maps of polarization-correlation maps of microscopic images for the healthy and diabetic kidney samples :, shown in Table 3.18 revealed the following differences:

Table 3.17 Quantitative characteristics of a probabilistic, coordinate, and scale-like structure of SCP modulus maps of healthy and pathologically changed rat kidney

Parameters	$|Sv_{i=3}(\Delta x, \Delta y)|$		$|Sv_{i=4}(\Delta x, \Delta y)|$	
Condition	Norm ($n = 39$)	Diabetes ($n = 39$)	Norm ($n = 39$)	Diabetes ($n = 39$)
Z_1	0.027 ± 0.0019	0.016 ± 0.0011	0.74 ± 0.034	0.089 ± 0.004
Z_2	0.013 ± 0.0011	0.007 ± 0.0005	0.23 ± 0.016	0.081 ± 0.009
Z_3	1.35 ± 0.12	3.78 ± 0.24	0.69 ± 0.055	2.07 ± 0.19
Z_4	3.24 ± 0.24	7.18 ± 0.39	0.82 ± 0.069	3.95 ± 0.22
Z_2^k	0.07 ± 0.005	0.11 ± 0.009	0.06 ± 0.005	0.09 ± 0.007
Z_4^k	2.48 ± 0.16	0.98 ± 0.078	1.63 ± 0.12	0.87 ± 0.068
D^f	0.22 ± 0.017	0.17 ± 0.013	0.32 ± 0.021	0.21 ± 0.017

Table 3.18 Quantitative characteristics of a probabilistic, coordinate, and scale-like structure of SCP phase maps of healthy and pathologically changed rat kidney

Parameters	$Arg(Sv_{i=3}(\Delta x; \Delta y))$		$Arg(Sv_{i=4}(\Delta x; \Delta y))$	
Condition	Norm ($n = 39$)	Diabetes ($n = 39$)	Norm ($n = 39$)	Diabetes ($n = 39$)
Z_1	0.019 ± 0.0014	0.013 ± 0.0011	0.92 ± 0.052	0.16 ± 0.012
Z_2	0.13 ± 0.011	0.047 ± 0.0033	0.15 ± 0.014	0.077 ± 0.003
Z_3	0.63 ± 0.052	1.97 ± 0.14	0.57 ± 0.046	1.77 ± 0.14
Z_4	0.79 ± 0.062	4.11 ± 0.35	0.68 ± 0.053	3.25 ± 0.27
Z_2^k	0.07 ± 0.004	0.12 ± 0.0092	0.054 ± 0.0055	0.072 ± 0.007
Z_4^k	3.65 ± 0.14	1.49 ± 0.079	4.64 ± 0.15	2.32 ± 0.064
D^f	0.25 ± 0.016	0.16 ± 0.014	0.26 ± 0.024	0.18 ± 0.014

- Statistical moments of the 1st–4th orders—$\Delta Z_1 = 1.51$–6.06 times; $\Delta Z_2 = 2.1$–2.6 times; $\Delta Z_3 = 3.04$–3.19 times; $\Delta Z_4 = 4.91$–5.24 times;
- Correlation moments of the 2nd and 4th orders—$\Delta Z_2^k = 1.4$–1.66 times; $\Delta Z_4^k = 1.95$–2.45 times;
- Dispersion of the logarithmic dependences of the power spectra of the distributions of the magnitude modulus of the 3rd and 4th parameters of the "two-point" Stokes vector of object fields $\Delta D^f = 1.47$–1.58 times.

3.7.1.2 Polarization-Correlation Maps of Samples of Rat Liver

In Figs. 3.18, 3.19, 3.20 and 3.21 shows 2D distributions ((1), (5), statistical distributions (2), (6),autocorrelation functions 2D distributions (3), (7), logarithmic dependences of power spectra 2D distributions (4), (8)) of SCP modulus and SCP phase of microscopic images of samples of a healthy ((1), (2), (3), (4)) and pathologically changed ((5), (6), (7), (8)) rat liver.

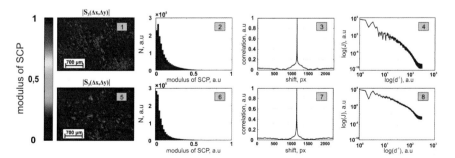

Fig. 3.18 Topographic (1), (5), statistical (2), (6), correlation (3), (7), fractal (4), (8) characteristics of SCP modulus $|Sv_{i=3}(\Delta x; \Delta y)|$ of a healthy ((1), (2), (3), (4)) and pathologically changed ((5), (6), (7), (8)) rat liver

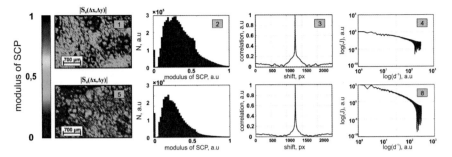

Fig. 3.19 Topographic (1), (5), statistical (2), (6), correlation (3), (7), fractal (4), (8) characteristics of SCP modulus $|Sv_{i=4}(\Delta x; \Delta y)|$ of a healthy ((1), (2), (3), (4)) and pathologically changed ((5), (6), (7), (8)) rat liver

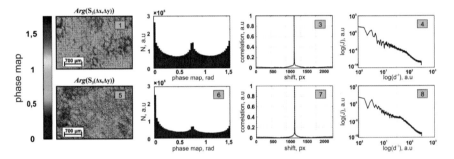

Fig. 3.20 Topographic (1), (5), statistical (2), (6), correlation (3), (7), fractal (4), (8) characteristics of SCP phase $Arg(Sv_{i=3}(\Delta x; \Delta y))$ of a healthy ((1), (2), (3), (4)) and pathologically changed ((5), (6), (7), (8)) rat liver

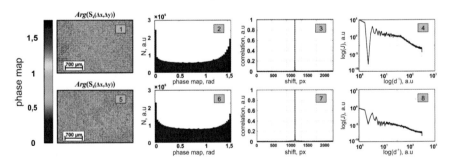

Fig. 3.21 Topographic (1), (5), statistical (2), (6), correlation (3), (7), fractal (4), (8) characteristics of SCP phase $Arg(Sv_{i=4}(\Delta x; \Delta y))$ of a healthy ((1), (2), (3), (4)) and pathologically changed ((5), (6), (7), (8)) rat liver

Comparison of the results of the polarization correlation technique of the distributions of SCP modulus $|Sv_{i=3;4}(\Delta x; \Delta y)|$ and phase $Arg(Sv_{i=3;4}(\Delta x; \Delta y))$ of polarization-inhomogeneous object fields liver samples of both healthy and diabetic types (see Figs. 3.18, 3.19, 3.20 and 3.21) with the data of mapping the distributions of parameters of the 3rd and 4th "two-point" parameters of the Stokes vector images of native preparations of kidney tissue (see Figs. 3.14, 3.15, 3.16 and 3.17) revealed similar patterns.

Quantitative differences between the statistical мометоами 1st–4th orders, correlation moments 2nd, 4th orders and fractal parameters (dispersion) describing the coordinate maps of SCP modulus and the phase of polarization-inhomogeneous images for liver samples are illustrated in Tables 3.19 and 3.20.

The obtained results of quantitative processing of data on the statistical, correlation, and scale-self-similar structure of maps of the module of polarization-inhomogeneous images for the healthy and diabetic liver samples: shown in Table 3.19 revealed the following differences:

Table 3.19 Quantitative characteristics of a probabilistic, coordinate, and scale-like structure of SCP modulus maps of healthy and pathologically changed rat liver

Parameters	$|Sv_{i=3}(\Delta x; \Delta y)|$		$|Sv_{i=4}(\Delta x; \Delta y)|$	
Condition	Norm ($n = 39$)	Diabetes ($n = 39$)	Norm ($n = 39$)	Diabetes ($n = 39$)
Z_1	0.074 ± 0.0045	0.057 ± 0.0033	0.37 ± 0.025	0.24 ± 0.016
Z_2	0.16 ± 0.013	0.12 ± 0.007	0.34 ± 0.014	0.23 ± 0.014
Z_3	0.64 ± 0.045	1.26 ± 0.062	0.47 ± 0.026	0.88 ± 0.052
Z_4	0.55 ± 0.037	1.43 ± 0.078	0.55 ± 0.037	0.74 ± 0.038
Z_2^k	0.067 ± 0.0055	0.075 ± 0.005	0.083 ± 0.0065	0.095 ± 0.0077
Z_4^k	2.59 ± 0.17	3.87 ± 0.23	1.16 ± 0.064	0.79 ± 0.058
D^f	0.23 ± 0.018	0.18 ± 0.016	0.34 ± 0.024	0.26 ± 0.014

Table 3.20 Quantitative characteristics of a probabilistic, coordinate, and scale-like structure of SCP phase maps of healthy and pathologically changed rat liver

Parameters	$Arg(Sv_{i=3}(\Delta x; \Delta y))$		$Arg(Sv_{i=4}(\Delta x; \Delta y))$	
Condition	Norm ($n = 39$)	Diabetes ($n = 39$)	Norm ($n = 39$)	Diabetes ($n = 39$)
Z_1	0.87 ± 0.046	0.68 ± 0.035	0.67 ± 0.042	0.54 ± 0.036
Z_2	1.15 ± 0.063	0.83 ± 0.058	1.05 ± 0.063	0.74 ± 0.044
Z_3	0.58 ± 0.035	0.86 ± 0.053	0.38 ± 0.023	0.59 ± 0.042
Z_4	0.88 ± 0.055	1.35 ± 0.072	0.63 ± 0.045	0.94 ± 0.053
Z_2^k	0.053 ± 0.0033	0.074 ± 0.008	0.029 ± 0.004	0.035 ± 0.0045
Z_4^k	2.79 ± 0.17	1.83 ± 0.23	3.08 ± 0.26	4.73 ± 0.35
D^f	0.28 ± 0.014	0.26 ± 0.017	0.24 ± 0.018	0.17 ± 0.0105

- Statistical moments of the 1st–4th orders ΔZ_1 = 1.26–1.44 times; ΔZ_2 = 1.36–1.52 times; ΔZ_3 = 1.81–1.95 times; ΔZ_4 = 1.41–2.63 times;
- Correlation moments of the 2nd and 4th orders—ΔZ_2^k = 1.05–1.14 times; ΔZ_4^k = 1.44–1.51 times;
- Dispersion of the logarithmic dependences of the power spectra of the distributions of the magnitude modulus of the 3rd and 4th parameters of the "two-point" Stokes vector of object fields ΔD^f = 1.29–1.37 times.

It was obtained for the SCP phase distributions:

- Statistical moments of the 1st–4th orders $\Delta Z_1 = \Delta Z_1$ = 1.23–1.31 times; ΔZ_2 = 1.41–1.44 times; ΔZ_3 = 1.41–1.57 times; ΔZ_4 = 1.54–1.56 times;
- Correlation moments of the 2nd and 4th orders—ΔZ_2^k = 1.2–1.41 times; ΔZ_4^k = 1.49–1.53 times;
- Dispersion of the logarithmic dependences of the power spectra of the distributions of the magnitude modulus of the 3rd and 4th parameters of the "two-point" Stokes vector of object fields ΔD^f = 1.29–1.47 times.

The results of the experimental measurement of polarization-correlation maps of topological distributions of the magnitude and phase of the 3rd and 4th parameters of the "two-point" Stokes vector of object fields revealed the most optimal conditions for the differentiation of changes in optical anisotropy as a result of pathology.

It has been established that the correlation matching of optical anisotropy parameters (optical axis directions, phase shifts) of the birefringent fibrillar networks of the kidney tissue is most pronounced—quantitatively the differences between the statistical, correlation and fractal moments of various orders that characterize the coordinate distributions of the module and phases of the 3rd and 4th parameters of the "two-point" Stokes vector of polarization-inhomogeneous object fields reach the same order.

For spatially unstructured, parenchymal tissues, the transformation of the statistical, coordinate, and scale-like structure of polarization-correlation maps is less pronounced. However, in this case too—significant sensitivity to the changes in optical anisotropy of the "islet" parenchymatous structures can be obtained—the values of the objective parameters has a 2-fold difference.

The next step was a comparative study of the diagnostic effectiveness of "single-point" (laser polarimetry) and "two-point" (Stokes-correlometry) methods of mapping phase-inhomogeneous object fields of samples of this type.

Tables 3.21 and 3.22 present the results of an information analysis of the diagnostic effectiveness of polarization correlometry, which are comparative with the laser polarimetry method.

The information analysis of the strength of "single-point" (laser polarimetry) and "two-point" (Stokes-correlometry) methods in differentiating poorly optically studied degenerative-dystrophic changes in the optically anisotropic component of biological tissues of rats of different morphological structures showed:

Table 3.21 Diagnostic power of the methods of polarizing correlometry and laser polarimetry of rat kidney tissue

Parameters	$Ac\,(\%)$					
	$|Sv_3|$	$|Sv_4|$	$ArgSv_3$	$ArgSv_4$	$\alpha(x,y)$	$\beta(x,y)$
Z_1	90.3*	92.5*	96.5*	97.4*	65.5	68.3
Z_2	91.2*	90.3*	94.4*	96.3*	64.6	67.6
Z_3	92.4*	87.8	97.2*	95.6*	82.7	83.2
Z_4	90,1*	91.6*	95.6*	98.7*	80.4	81.4
Z_2^k	82.3	81.4	85.4	87.5	62.6	64.7
Z_4^k	93.8*	93.5*	98.2*	98.4*	83.4	82.3
D^f	80.6	82.4	84.6	86.6	67.5	68.6

Table 3.22 Diagnostic power of the methods of polarizing correlometry and laser polarimetry of rat liver tissue

Parameters	$Ac\,(\%)$					
	$|Sv_3|$	$|Sv_4|$	$ArgSv_3$	$ArgSv_4$	$\alpha(x,y)$	$\beta(x,y)$
Z_1	85.4	83.2	92.5*	91.6*	63.5	65.4
Z_2	86.3	88.4	90.7*	92.7*	61.7	63.2
Z_3	84.6	82.3	91.6*	90.5*	79.3	78.5
Z_4	90.7*	90.6*	90.4*	92.8*	77.1	75.8
Z_2^k	78.5	77.5	80.7	81.3	58.3	59.5
Z_4^k	90.3*	90.8*	92.3*	92.6*	74.7	77.2
D^f	76.6	75.2	80.5	81.3	60.7	61.5

- Low (for kidney tissue $\sim 80\%$ and $\sim 75\%$ liver tissue) level of accuracy of laser polarimetry technique;
- Excellent level of accuracy for the methods of Stokes-correlometry of the structured $(94\% \leq \max Ac \leq 98\%)$ and parenchymatous $(90\% \leq \max Ac \leq 92\%)$ biological tissues.

References

1. Cassidy, L.: Basic concepts of statistical analysis for surgical research. J. Surg. Res. **128**(2), 199–206 (2005)
2. Davis, C.S.: Statistical methods of the analysis of repeated measurements. Springer, New York (2002)
3. Petrie, A., Sabin, C.: Medical Statistics at a Glance. Wiley, Chichester, UK (2009)
4. Ushenko, A., Pishak, V.: Laser polarimetry of biological tissue: principles and applications. In: Tuchin, V. (ed.) Handbook of Coherent-Domain Optical Methods: Biomedical Diagnostics, Environmental and Material Science, pp. 93–138 (2004)

5. Angelsky, O., Ushenko, A., Ushenko, Y., Pishak, V., Peresunko, A.: Statistical, correlation and topological approaches in diagnostics of the structure and physiological state of birefringent biological tissues. In: Handbook of Photonics for Biomedical Science, pp. 283–322 (2010)
6. Ushenko, Y., Boychuk, T., Bachynsky, V., Mincer, O.: Diagnostics of structure and physiological state of birefringent biological tissues: statistical, correlation and topological approaches. In: Tuchin, V. (ed.) Handbook of Coherent-Domain Optical Methods. Springer (2013)
7. Deboo, B., Sasian, J., Chipman, R.A.: Degree of polarization surfaces and maps for analysis of depolarization. Opt. Exp **12**, 4941–4958 (2004)

Chapter 4
Conclusions

1. Analytical relationships between the set of parameters of the "two-point" Stokes vector and the directions of the optical axes and phase shifts of the fibrillar networks of polycrystalline components of biological tissues of different morphological structure and pathological state are determined.
2. A polarization-correlation mapping technique for the distributions of the magnitude and phase parameters of the "two-point" Stokes vector of phase-inhomogeneous object fields of optically anisotropic layers of biological tissues has been developed and analytically justified.
3. For partially depolarizing biological layers based on the layer-by-layer Mueller-matrix description of the parameters of optical anisotropy, algorithms for the experimental determination of the elements of differential matrices of the 1st and 2nd orders are determined.
4. Algorithms for polarization-phase tomography are found in reconstructing the distributions of the mean linear and circular birefringence and dichroism of the polycrystalline component of partially depolarizing layers of biological tissues.
5. Algorithms of diffuse tomography in reconstructing the distribution of fluctuations of linear and circular birefringence and dichroism of the polycrystalline component of the depolarizing layers of biological tissues are determined.
6. In the framework of the statistical approach to the analysis of the results of polarization optical techniques for reconstructing the distributions of average and fluctuations of the phase and amplitude anisotropy parameters, the criteria for reliable differentiation of necrotic changes in the structure of fibrillar networks are determined of the myocardium (type A and type B—excellent quality), as well as the vaginal wall with prolapse of the genitals (type C and type D—good quality), early stages of endometrial cancer of the uterine wall (good quality).

V. T. Bachinskyi et al., *Polarization Correlometry of Scattering Biological Tissues and Fluids*, SpringerBriefs in Physics, https://doi.org/10.1007/978-981-15-2628-2_4

7. Data were obtained on the effectiveness of using Mueller-matrix mapping of depolarization maps of model diffuse layers—phantoms with different scattering coefficients for realizing the perspectives of using this optical technique in clinical application for diagnosis of cancerous tissues.

8. Mueller-matrix mapping of depolarization maps is effectively tested using of statistical, correlation and fractal analysis for the analysis of paraffin-embedded tissue blocks one can unambiguously detect the malignant transformation of tissue.

9. The method of 3D Mueller-matrix diffuse tomography of distributions $DT_{red}\{B\tilde{L}_{0;90}; B\tilde{L}_{45;135}; B\tilde{C}_{\otimes;\oplus}\}(\varphi_1 = 0.3\,\text{rad}; 2\varphi_1; , \ldots, 6\varphi_1)$ of fluctuations in the parameters of linear and circular birefringence of partially depolarizing polycrystalline films of biological fluids has been theoretically substantiated and experimentally tested. The transformation scenario the magnitude of the set of statistical moments of the first–fourth orders characterizing layer-by-layer maps of the fluctuations of the phase anisotropy of a polycrystalline film of blood of a healthy donor in different "phase" sections of its volume has been examined and analyzed. The most sensitive to prostate cancer parameters are statistical moments of the third and fourth orders $(\Delta X^*_{i=3;4} \equiv \Delta X_{i=3;4}(\varphi^* = 0.85\,\text{rad}) \rightarrow \text{max})$, which characterize the distribution of fluctuations $B\tilde{L}_{0;90}; B\tilde{L}_{45;135}; B\tilde{C}_{\otimes;\oplus}$ in the parameters of phase anisotropy of polycrystalline blood films of practically healthy donors and patients with prostate cancer. Respectively, good $(80\% \leq \text{Ac}(X_{3;4}(B\tilde{L}_{0;90;45;135})) \leq 85\%)$ and excellent $(\text{Ac}(X_{3;4}(B\tilde{C}_{\otimes;\oplus})) \geq 90\%)$ accuracy of the 3D Mueller-matrix tomography approach for differentiating samples of whole-blood polycrystalline films of healthy donors and patients with prostate cancer has been achieved.

10. The new polarization-correlation mapping technique of the distributions of the величины modulus and phase of "two-point" Stokes-vector parameters of object fields of biological tissues with different morphological structures and physiological states has been introduced. The obtained polarization-correlation maps of microscopic images of normal and diabetic samples rat kidney and liver were examined.

11. A comparative analysis of the diagnostic effectiveness of the "one-point" and "two-point" methods of polarization and correlation mapping using the off statistical, correlation and fractal analysis demonstrated the excellent accuracy $(Ac \geq 90\%)$ in diagnosis of variations in optical anisotropy of biotissues by using the Stokes-correlometry approach.

Printed in the United States
By Bookmasters